Astronomers' Universe

For further volumes:
http://www.springer.com/series/6960

Chris Kitchin

Exoplanets

Finding, Exploring, and Understanding
Alien Worlds

 Springer

Chris Kitchin

ISBN 978-1-4614-0643-3   e-ISBN 978-1-4614-0644-0
DOI 10.1007/978-1-4614-0644-0
Springer New York Dordrecht Heidelberg London

Library of Congress Control Number: 2011937483

Printed on acid-free paper

Springer is part of Springer Science+Business Media (www.springer.com)

For Christine

Preface

Alien worlds, extra-solar planets, deep space planets, exoplanets – whatsoever we choose to call them, what sort of objects do we mean when we talk of planets belonging to stars other than the Sun?

A few years ago the TV, radio and newspapers were full of the temerity of astronomers who had demoted Pluto from being one of the nine major planets in the solar system to something called a 'Dwarf Planet'. In 2006 the International Astronomical Union (IAU) made this decision because of the discovery of solar system objects further from the Sun than Pluto that were similar in size to Pluto. One of these, Eris, is actually larger than Pluto and was briefly called the tenth planet of the solar system. It was the prospect of many more such objects being found and the number of planets becoming unmanageable that led the IAU to change Pluto's status. However the IAU has no legal standing and many professional and most amateur astronomers do not belong to it. Thus anyone who still wishes to regard Pluto as the ninth planet of the solar system is perfectly entitled to do so.

When it comes to planets beyond the solar system the IAU has no official definition – indeed the details of the existing classification actually mean that the word 'planet' *only* applies to eight objects within the solar system. Unofficially a number of varied criteria are in use to define an 'exoplanet'. Most definitions agree that if the object's mass is more than thirteen times the mass of Jupiter then it is too big to be called an exoplanet but should be classed as a type of 'failed star' known as a brown dwarf.

For those objects below the 13 Jupiter-mass limit though:

Are satellites to be included?
Are objects as small as our Moon to be included?
What of objects orbiting brown dwarfs?
and
What of objects that float free of any star by themselves in space?

In these areas opinions vary regarding which objects should be called exoplanets and which should be classed as something else.

Names are useful shorthand labels, but should not dominate the subject as the recent debate over whether Pluto is a planet or a dwarf planet has done. Lewis Carroll has his own take on the importance of names and other words:

> " 'When *I* use a word,' Humpty Dumpty said in rather a scornful tone, 'it means just what I choose it to mean – neither more nor less.'
> 'The question is,' said Alice, 'whether you CAN make words mean so many different things.'
> 'The question is,' said Humpty Dumpty, 'which is to be master – that's all.'"

Through the Looking Glass

Acting on Dumpty's principle a broad definition will be adopted in this book. The term 'Exoplanets' will cover objects ranging from small asteroids (say 1,000 m across or a mass 0.00000000001% that of the Earth) to just short of the failed stars known as Brown Dwarfs (4,000 Earth masses, 13 Jupiter masses). Of course sometimes sub-divisions, such as Planetesimals, Super Earths, Hot Jupiters, etc. will prove to be useful and objects outside the defined range of exoplanets, such as dust particles and small stars will come into the discussions at times. This book though is mostly about the menagerie of sub-stellar entities, whatever they may be called and whenever, howsoever and wherever they are to be found in the universe.

Our Sun is a pretty commonplace star and, as we well know, it is accompanied by a host of planets, dwarf planets, asteroids, satellites, comets and the like, each gravitating around the Sun and themselves in a complex and un-repeating 4,500 million year long ballet.

If the Sun is a typical star, then surely other stars must also have their retinues of planets and satellites? By the late twentieth century many astronomers were beginning to think that the Sun's planetary family must be a rare and unusual occurrence because decades of searching for planets beyond the solar system had failed to turn up any examples.

The situation changed abruptly in the 1990s. Firstly in 1992 Aleksander Wolszczan and Dale Frail discovered two rocky planets orbiting the pulsar PSR B1257+12 (see Appendix I for an explanation of stars' and exoplanets' names and labels). Then in 1995 came the real break-through when Michel Mayor and Didier Queloz found the first planet belonging to a normal star. From a good observing site, that star, named 51 Pegasi, may just be seen with the naked eye about halfway between the bright western stars of the square of Pegasus. 51 Peg is very similar to our Sun though a bit older and its planet has a mass about half that of Jupiter. What came as a major surprise in 1995 however was that the planet is only 7,500,000 km away from its host star – an eighth of Mercury's distance from the Sun. The exoplanet's surface temperature reaches 1,200°C – hot enough to melt most rocks. 51 Peg's planet though is a gas giant like Jupiter.

Writing in early 2011, we know of around 530 exoplanets, many of which are giant planets close in to their stars like 51 Peg's planet. Sufficient is now understood about exoplanets that we are no longer restricted just to examining individual planets but we may begin to develop ideas and come to conclusions about the properties, natures and characteristics of planets that have a broader application and validity throughout the universe.

For the first time in the history of human science we may begin to see the importance of the Earth and the solar system within a wider context and not just as the local neighbourhood wherein we happen to live. The aims of this book are thus

- To conduct the reader through the heady experience of exploring one of the most exciting and rapid establishments of a new area of science that has ever happened,
- To explore the avalanche of dramatic discoveries of new planets that have been made over the last decade-and-a-half,
- To seek out how and why those discoveries have been made possible and to highlight where amateur astronomers can contribute to the research,
- To probe what we now know about exoplanets – both for individual planets and the more universally applicable trends,

and last, but not least,

- To investigate whether or not we might ever travel to and perhaps colonize an exoplanet.

I have assumed that the reader will have some prior knowledge of astronomy but not beyond the level of a well-read person who has an interest in the sciences generally. If you do find something that is unfamiliar and need to look it up, then a recently published introductory astronomy book, an astronomy dictionary or the internet should be sufficient and a list of suggested sources of other reading is provided at the end of this book. For those of you who wish to know more, deeper briefings about some of the technicalities behind finding, exploring and understanding alien worlds are also included at the end of the book. BUT – you do not need to read those sections or deal with equations in order to enjoy the main part of the book and to see how scientists really get to work in a brand new theatre of science.

I hope that you are pleased with the book and find it interesting and useful.

Happy Reading!

Hertford   Chris Kitchin

About the Author

Chris Kitchin has written or contributed to over two dozen books, and has published more than 500 articles in the astronomical journals and magazines. He also appears regularly on television, including many appearances on BBC TV's *Sky at Night*. His works for Springer includes, *A Photo Guide to the Constellations: A Self-Teaching Guide to Finding Your Way Around the Heavens* (1997), *Solar Observing Techniques* (2001), *Illustrated Dictionary of Practical Astronomy* (2002), and most recently *Galaxies in Turmoil* (2007). In his 'day job' Chris is Emeritus Professor of Astronomy at the University of Hertfordshire, where until ten years ago he was also Head of Physics and Astronomy, and Director of the University Observatory. Like many other astronomers Chris's interest in the subject started early. At the age of fourteen, he constructed an 8-inch Newtonian after spending hundreds of hours grinding and polishing the main mirror from scratch. Despite using some of the largest telescopes in the world since then, Chris still enjoys just 'gazing at the heavens' - though nowadays it's through a Zeiss Maksutov telescope.

Acknowledgements

This book could only be written because of the hard work of an enormous number people. Many of them are referred to by name within the text and to them and to all the others for whom there was insufficient room to give the proper credit – my many thanks and my best wishes in all your future endeavours – may you all discover the exoplanets of your dreams!

Especial thanks are owing to Professors Dale Frail, Geoff Marcy and Didier Queloz for their patience in responding to my questions so fully and helpfully. Last and very definitely not least my thanks go to my wife Christine and to John Watson for their detailed and perceptive comments on my draft text which have improved it superlatively. The remaining deficiencies are exclusively my own.

Chris Kitchin

Contents

1. Because We Live on One! – or – Why Planets and Exoplanets Are Important

Because *we* live on a planet, planets other than the Earth are fascinating, significant and important to us, whether they form part of the solar system, belong to stars other than the Sun, or even float freely in space independent of any star. As well as a purely intellectual interest in planets and exoplanets, there is also the hope that one day humans might set up colonies on some of them. Thus providing a safety net against the remote chance of human kind being wiped-out by a large meteorite impact with the Earth or the far more likely possibility that we shall render the Earth unfit for life ourselves.

While any exoplanet is better than none, if we are truly honest with ourselves then our interest is even more parochial than that – what we *really* want to find are exo-Earths.

Exo-Earths are planets inhabitable by human-kind. A twin-Earth – immediately ready for us to live on would be best of all – but they will be very few and far between. Most people would probably settle for an exo-Earth that was 75% or 80% of the ideal.

Thus the holy-grail of exoplanet-hunting teams is currently to find the 'little-blue-dot' (see Chap. 2) that would mark the discovery of an exo-Earth and missions such as 'Kepler' have this amongst their primary aims. Our requirements for an exo-Earth would include a reasonable gravitational pull (we wouldn't want to weigh half a ton or alternatively to risk floating off into space when attempting a high jump), a comfortable temperature and a breathable atmosphere. Because an exo-Earth must be 'just right' for ourselves, like the temperature of Baby Bear's porridge and the softness of his bed in the tale of *Goldilocks and the Three Bears*, they are also often referred to as 'Goldilocks Planets.'

C. Kitchin, *Exoplanets: Finding, Exploring, and Understanding Alien Worlds*, Astronomers' Universe, DOI 10.1007/978-1-4614-0644-0_1,
© Springer Science+Business Media, LLC 2012

We have yet to find a true Goldilocks exoplanet. The nearest approach so far is an exoplanetary candidate (i.e. a star that has been observed to change brightness in a way that *might* arise from the transit of an exoplanet across its disk, but for which the reality of the exoplanet remains to be confirmed – see Chap. 6) observed by the Kepler spacecraft in 2009 and labelled KOI 326.01 (Kepler Object of Interest). The star involved is on the Cygnus/Lyra border and is a cool red dwarf. If the star's exoplanet is confirmed and the preliminary estimates of the planet's properties are at least roughly correct, then the planet is around seven to eight million kilometres out from its host star and is a little smaller than the Earth. If the planet does not have an atmosphere then its surface temperature is likely to be around 60°C. If there is an atmosphere then the surface temperature is probably somewhat higher than this value. 60°C or a bit higher is uncomfortably warm for humans but there are numerous terrestrial organisms that can flourish in temperatures up to 100°C (Chap. 14). KOI 326.01 is thus the first exoplanet or potential exoplanet found that is of about the Earth's size and which is orbiting within its star's habitable zone (the habitable zone is the region surrounding a star within which liquid water can potentially exist – Chap. 14). The announcement of the detection of KOI 326.01 occurred during the final stages of writing this book (early 2011) and the star will undoubtedly be the centre of an intensive observing campaign from this time onwards so confirmation or otherwise of its exoplanet should occur fairly quickly.

A remarkable planetary system surrounds Gliese 581 (Figure 1.1a), a faint cool star in Libra that lies just 20 light years away from us. Six exoplanets are thought to orbit the star, although the two most recent discoveries remain to be confirmed. Three of the planets verge on being habitable. The very recently discovered (and unconfirmed) exoplanet, Gliese 581 g could have an average surface temperature between –30°C and –10°C, which would make conditions comparable with the fringes of Antarctica. It seems possible though that the planet has a reasonably dense atmosphere so the average temperature could be higher than these values. Furthermore the planet always keeps the same face towards its star (like our Moon always keeps the same face towards the Earth) so the 'day' side will have a much higher temperature than the

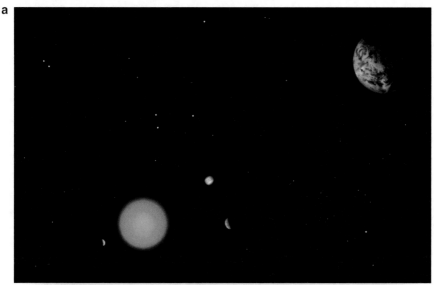

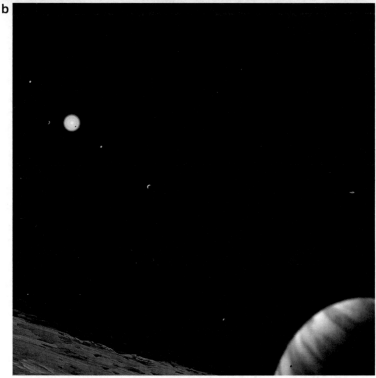

FIGURE 1.1 (a) An artist's impression of the star Gliese 581 with the (uncon-firmed) exoplanet Gliese 581 g in the foreground. Three other planets of this six-planet system are shown in the distance. (b) An artist's impres-sion of the star Kepler-11 seen from near an hypothetical satellite of the outermost planet, Kepler-11 g. The five inner planets are shown, with one in transit across the star's disk. A second (also hypothetical) satellite of Kepler-11 g is included which is casting a shadow onto its planet. (Copy right © C.R. Kitchin 2010).

average and the 'night' side a much lower one. In the twilight zone between the 'day' and 'night' sides of the planet regions should exist that have ideal temperatures for humans. Gliese 581 g has about three times the mass of the Earth and if it is rocky like the Earth, its surface gravity will be about 40% higher than that of the Earth – a 70 kg human would weigh 100 kg on the planet.

In addition to KOI 326.01, in early 2011 the Kepler team also announced another 53 exoplanetary candidates that were in or near the habitable zones of the stars. Four of these are super-Earths (i.e. with radii probably less than twice that of the Earth) while the remainder are comparable with Neptune or Jupiter in size or even larger. Super-Earths could potentially be inhabitable by humans, although other things being equal, their surface gravities would be up to twice that of the Earth. Planets of the size of Neptune or Jupiter are probably not inhabitable by humankind though other life forms might well be able to exist upon them. However any satellites of such planets will also be within the star's habitable zone and the larger such objects could have atmospheres (like Saturn's Titan) thus potentially providing the conditions required for the existence of life.

No exoplanets at all are known with oxygen-rich atmospheres. Thus a true twin-Earth remains to be found at the time of writing, but there can be little doubt that success in that search is only a matter of persistence. Of course, when we do find an exo-Earth or a twin-Earth we may also find it to be inhabited by ETs (intelligent Extra-Terrestrials or alien creatures) – further speculation on that topic however is left for later on in this book.

In addition to having a parochial interest in discovering a twin-Earth, many people would probably also like to know whether or not our whole solar system has any look-alikes. The multi-planet system Gliese 581 has already been mentioned, but a simple inspection of the 600 or so known exoplanets would suggest that the answer to that query must be 'No.' From the first exoplanet found around a normal star in 1995 onwards, the majority of exoplanets have been found to huddle very near to their host stars – often very much closer to their stars than even Mercury is to the Sun. Many of these close-in exoplanets are also gas giants as large as or larger than Jupiter. Furthermore, three quarters of currently detected exoplanets are singletons – i.e. the only exoplanet known for their host star.

There are though over 50 known multi-exoplanet systems with up to six confirmed planets in a single system, so clearly the solar system is not unique in possessing many planets. However these exoplanetary systems also huddle close to their stars – the very recently discovered Kepler-11 for example has six exoplanets that are all closer to their star than Venus is to the Sun (Figure 1.1b). Only 10% of the multi-exoplanet systems have planets as far out or further out from their host stars as Jupiter is from the Sun.

The exoplanetary system 47 UMa is the closest analogue to the solar system that has been found to date. The star has three known exoplanets with masses from half to two-and-a-half times that of Jupiter that are in orbits ranging from 2 to 11 astronomical units (see Appendix II for a note on the units used in this book) in radius. The most far flung multi-exoplanet system is currently HR 8799 which has four exoplanets that are all much more massive than Jupiter and which are in orbits ranging from 14.5 to 68 astronomical units out from their star.

Thus the currently known multi exoplanet systems do not resemble the solar system much at all. However that is not the final word on the matter. Almost all the methods used to detect exoplanets (Chaps. 5–10) have a predilection for detecting massive planets orbiting close in to their host stars. Detecting smaller planets in larger orbits is much more difficult. Our results at the moment are therefore dominated by compact exoplanetary systems and massive planets. Almost certainly more extensive systems containing both small and large exoplanets do exist but have yet to be found. To return to the question of whether or not there are twin solar systems somewhere out there, the true answer is probably 'Yes' – but the next generation of exoplanet detectors will be needed to find them.

In contrast to our rather obsessive interest in and assessment of the importance of exoplanets, aliens who do not dwell on one would probably think that they are of little significance – and their viewpoint is likely to be the more realistic one. The major and spectacular sights and features within the universe include the glories of stars and galaxies, the magnificence of gaseous nebulae and dust clouds and the dramatic convulsions of supernovae and gamma ray bursters. Exoplanets by contrast are extremely difficult to find, are unspectacular and are un-photogenic. They also form only a very tiny fraction of the mass of the universe – perhaps less than 0.01%. In the general scheme of things exoplanets are thus not a particularly important component of the universe. ETs living, say, on stars or within giant molecular clouds would undoubtedly regard exoplanets as having only a very minor and peripheral relevance to their attempts to understand the nature of the universe.

However, whilst the author will be grateful for any sales of this book in the ET market, it *is* written for humankind and so the remaining chapters will be devoted to attempting to satisfy our curiosity about the matter.

2. A Quick Tour of the Exoplanet Menagerie

Alien planets come in several varieties – some types we know and love from looking around the solar system, others are very different from anything we have previously encountered. The main groups of exoplanets found so far (doubtless others will turn up in due course) are Hot Jupiters, Hot Neptunes, Cold Jupiters, Super Jupiters, Super Earths, Little Blue Dots (or Exo-Earths, Twin-Earths) and Free-Floating Planets.

Hot Jupiters

Three-quarters of the exoplanets found so far have masses within the range a half to 13 times that of Jupiter. Over 40% of these planets are closer to their host stars than the Earth is to the Sun. Their proximity to their host stars means that such exoplanets have very high cloud top temperatures and since they are also very massive, they have become known as hot Jupiters. WASP-19b (see Appendix I for an explanation of exoplanet names), for example, orbits a mere 1,800,000 km above its star's surface – just 3.5 times the distance of the Moon from the Earth.

There is no reason to think that 40% of *all* exoplanets are hot Jupiters. In fact it is probable that only a small fraction of exoplanets are hot Jupiters, although it is likely that they will predominate in terms of mass. The underlying cause of the high proportion of hot Jupiters in the current sample of exoplanets is that they are simply the easiest exoplanets to find.

Although the cloud top temperatures of hot Jupiters can be 2,000°C or more, and they are predominantly made up from the lightest gases – hydrogen and helium – those gases will not boil off. The gravitational fields of such massive objects are sufficient

C. Kitchin, *Exoplanets: Finding, Exploring, and Understanding Alien Worlds*, Astronomers' Universe, DOI 10.1007/978-1-4614-0644-0_2, © Springer Science+Business Media, LLC 2012

to hold on even to hydrogen at temperatures of several thousand degrees. The high temperature will, though, lead to the exoplanet being bloated in size in comparison with our 'own' Jupiter. WASP-17b, for example, has half Jupiter's mass, but 1.8 times Jupiter's radius, giving it an average density about the same as that of expanded polystyrene foam!

Hot Neptunes

A small group of exoplanets that are similar to hot Jupiters, but with lower masses. The minimum mass to retain a substantial hydrogen atmosphere is around 3% that of Jupiter (ten times the mass of the Earth). The transition point between hot Neptunes and hot Jupiters is fairly arbitrary, but a mass of a fifth that of Jupiter is sometimes used.

Cold Jupiters

About a third of the massive exoplanets discovered to date are at least twice as far from their stars as the Earth is from the Sun. Since the host stars are often small (because this also makes their exoplanets easier to find) and are therefore relatively cool, their exoplanets have cloud top temperatures comparable with that of Jupiter (around –140°C). Hence by analogy with hot Jupiters, this class of exoplanets is called the cold Jupiters or sometimes twin Jupiters. Cold Jupiters may well resemble Jupiter itself in appearance (Figure 2.1), especially if they rotate relatively quickly (Jupiter's day is just 10 h long).

The first cold Jupiter, 55 Cnc d, was found in 2002 by Geoff Marcy and Paul Butler. 55 Cnc d has a mass four times or more that of Jupiter and orbits a solar-type star, 55 Cnc A (also known as ρ Cnc – the Greek alphabet is listed in Appendix I for reference) some 40 light years away from us. 55 Cnc A has at least four other exoplanets and may also form a binary system with the star, 55 Cnc B, a red dwarf that is 1,100 astronomical units away from the main star. The cold Jupiter exoplanet has a 14-year orbital period

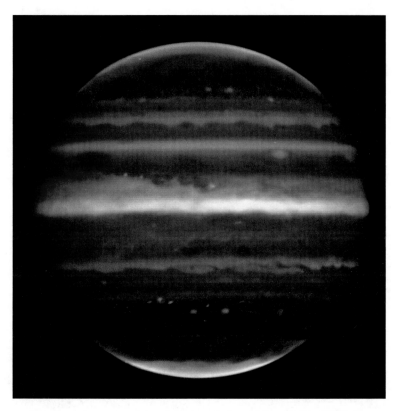

FIGURE 2.1 Jupiter imaged in the near infrared by ESO's VLT. Some, perhaps many, cold Jupiters' appearances may resemble this. (Reproduced by kind permission of ESO, F. Marchis, M. Wong, E. Marchetti, P. Amico and S. Tordo).

around 55 Cnc A and is around 5.8 astronomical units out from the star (cf. Jupiter's 11.9 years and 5.2 astronomical units).

Super Jupiters

Exoplanets with masses five times that of Jupiter or more are sometimes put together as a group called Super Jupiters (or Mega-Jupiters). The upper limit for super-Jupiters should be 13 Jupiter masses (the transition mass to brown dwarfs), but higher mass objects, which may be genuine exoplanets or small brown dwarfs, are sometimes included within this grouping.

Super Earths

Exoplanets just a little more massive than the Earth – say from 1.5 up to 10 Earth masses (3% of Jupiter's mass) – are classed as Super-Earths. Because the quoted exoplanet masses are usually the minimum possible values it is likely that some of the super-Earths at the top end of this range actually exceed the ten Earth mass limit. The lowest mass super-Earth currently known is Gliese 581e at just under two Earth masses.

The smaller super Earths are likely to resemble the Earth in being largely rocky in composition. Whether or not a super Earth has an atmosphere will depend upon its evolutionary history and its proximity to its host star (and hence its surface temperature – too high a temperature and the atmosphere will boil away).

Exo-Earths, Goldilocks Planets, Twin Earths and Little Blue Dots

Exoplanets with masses less than about one and a half times that of the Earth are called exo-Earths whether they are close to their host star or further out. No confirmed exoplanet with a mass as small as that of the Earth (except for PSR 1257+12 b at 2% of the Earth's mass – see later) has been found at the time of writing.

Goldilocks planets are exo-Earths that have orbits placing them at sufficient distances from their host stars that liquid water could potentially exist upon them. This requires temperatures in at least some places on or within the planet to be in the region of 0°C to 100°C+. The region around the star where such planetary temperatures are possible is termed the habitability zone since we expect life as we know it to require liquid water. Determining the whereabouts of the habitability zone however is not simple since factors such as whether the planet has an atmosphere or not, whether the planet rotates with respect to its host star or always keeps the same face towards it, whether the planet has internal heat sources (volcanoes) and so on come into play. The Kepler spacecraft though has recently observed an object that may be a

Goldilocks exoplanet but which has yet to be confirmed even to be a planet – the object could be something else such as an eclipsing binary star. If the object is confirmed to be an exoplanet then KOI 326.01 (Kepler Object of Interest) could be around 80% of the Earth's mass. Furthermore, although it is just seven or eight million kilometres out from its host star, that star is a faint red dwarf and so the planet's temperature could be as low as 60°C placing it firmly within the habitability zone.

Twin Earths and Little Blue Dots are Goldilocks planets that additionally have most or all of the other requirements for humans to live on them (Little Blue Dots are so called because if we ever find a twin-Earth and could build some sort of telescope capable of imaging it directly a 'Little Blue Dot' is exactly how it would appear). Primarily this would mean an oxygen-rich atmosphere but there would be a myriad of other requirements. Whether or not KOI 326.01 might be a twin Earth remains to be seen, but the odds are against it. The Kepler spacecraft may detect one or more examples (without obtaining direct images) before its mission concludes sometime between 2013 and 2016.

Free-Floating Planets

Some exoplanets have been found that are not associated with host stars but which float as independent entities within the galaxy. A couple of dozen or so of these objects have been detected to date, many within the Orion nebula (M 42). At the top end of the mass range, free-floating planets blend into the smaller free-floating brown dwarfs. Some astronomers argue that a planet (or exoplanet) has to 'belong' to a host star. As discussed earlier though free-floating planets are considered here to be *bona fide* exoplanets – not least because some of them will have been formed within a star's planetary system and subsequently ejected during gravitational interactions with other planets. Synonyms for free-floating planets include – Inter-stellar planet, Inter-stellar comet, Isolated Planetary Mass Object (IPMO), Orphan planet, Planemo, Planetar, Rogue planet and Sub-brown dwarf.

Just How Many Exoplanets Are There?

Many of the methods of detecting exoplanets have intrinsic biases, especially towards finding hot Jupiters. The currently observation that 40% of all known exoplanets are hot Jupiters or hot super-Jupiters is thus probably a large over-estimate. It is thus still early days to give any reasonably reliable estimates of exoplanet numbers. Nonetheless a number of indicators suggest that they occur frequently.

At the time of writing, the results of observations by the Kepler spacecraft are only available up to February 2011. Furthermore Kepler only observes about 0.25% of the whole sky and concentrates on solar type stars out to a distance of 3,000 light years away from us. Nonetheless in excess of 1,600 exoplanetary candidates have been found by the mission. The Kepler team estimate that around 80% of their exoplanetary candidates will eventually be confirmed to be true exoplanets, suggesting that at least a 100 million exoplanets are out there somewhere within the Milky Way galaxy.

Recent Keck telescope observations of 166 Sun-like stars (spectral types G and K – see Appendix IV for a brief summary of stellar spectral and luminosity classification) by Andrew Howard and Geoff Marcy suggest that for these star-types 13% of the stars have one or more exoplanets. Their results predict that about 1–2% of Sun-like stars have Jupiter-sized planets, 6% have Neptune-sized planets and 12% have super-Earths. Extrapolating from this data indicates that up to 23% of such stars may have Earth-mass exoplanets. If correct, this would suggest that the Milky Way galaxy might be home to between 10,000 million and 20,000 million exo-Earths and around twice that number of exoplanets would be of the Earth's mass or more.

However lower mass stars (spectral type M) are far more numerous than solar-type stars within the Milky Way – in the solar locality, for example, red dwarfs comprise three out of every four stars. If similar proportions of red dwarf stars have exoplanets then the galactic population could be up to 200,000 million. If we extend the extrapolation down to the mass of Mercury (the least massive planet within the solar system) then the numbers could be ten to a hundred times higher still. Thus a 'ball-park' figure for the number of exoplanets within the Milky Way galaxy of up to 10,000,000 million is a possibility – more than the number of stars in the galaxy. Observations in 2010, again using the Keck telescopes, have shown that red dwarfs may be up to 20 times more abundant relative to solar-type stars in elliptical galaxies than they are in our own, so that a large elliptical galaxy could perhaps be home to 1,000,000,000 million exoplanets.

3. An Exoplanet Retrospective

This is admittedly a personal and idiosyncratic journey through the history of discoveries and events, facts and ideas, possibilities, wild speculations and pure inventions relating to our understanding of planets, exoplanets, life and alien life. Nonetheless it should enable the reader to appreciate that the origins of the very recent, modern and rapidly developing study of planets beyond the solar system actually has its roots extending back centuries if not millennia.

Year	Discovery, event, fact, idea, possibility or wild speculation, etc.
Pre-recorded history	Someone notices that although most objects in the sky keep the same relative positions, a few move around. Five star-like heavenly bodies (or occasionally seven because the dawn and evening apparitions of Mercury and Venus were sometimes thought to be different objects) are identified that move regularly with respect to the thousands of fixed objects in the sky. The name, planets, given to these moving bodies derives from the Greek *asters planetai* (wandering stars). We now name these planets Mercury, Venus, Mars, Jupiter and Saturn
c. 530 BC	Pythagoras (c. 570–c. 495 BC) proposes the idea of a spherical Earth, although he may have adopted the concept from earlier Greek philosophers
c. 400 BC	Philolaus (c. 470–c. 385 BC) suggests that the Earth, planets, Sun, Moon and an anti- or counter-earth (also known as the Antichthon) all move around a 'central fire' (Figure 3.1)
c. 380 BC	Plato (428 or 427–348 or 347 BC) teaches that all celestial objects move along perfect circles at uniform speeds – an idea that will bedevil astronomy for two millennia. The resulting models for the solar system had to contain numerous deferents, epicycles, eccentrics and equants in order to account for the observed non-uniform movements. Ptolemy's model of the solar system, for example, contained 80 circles in order to give a reasonably accurate prediction of the positions of the planets. Even Copernicus' heliocentric model retained 34 such circles. Not until Kepler introduced the concept of elliptical orbits in 1609 was Plato's dictum finally refuted

(continued)

C. Kitchin, *Exoplanets: Finding, Exploring, and Understanding Alien Worlds*, Astronomers' Universe, DOI 10.1007/978-1-4614-0644-0_3,
© Springer Science+Business Media, LLC 2012

(continued)

Year	Discovery, event, fact, idea, possibility or wild speculation, etc.

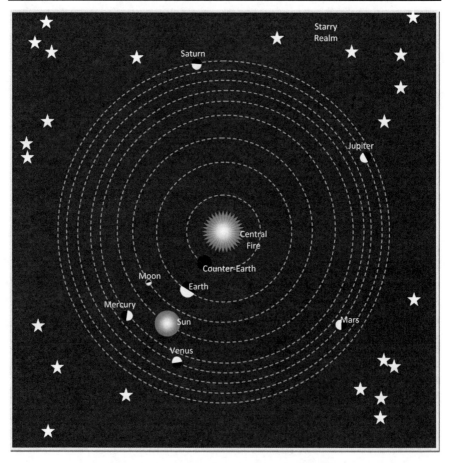

FIGURE 3.1 Philolaus' central-fire model of the solar system. This is the first suggestion that the Earth is a planet and is moving through space (although not around the Sun).

c. 350 BC	Heracleides of Pontus (c. 390–c. 310 BC) suggests that the heaven's apparent rotation is actually due to the counter rotation of the Earth. He may also have proposed a Sun-centred model for the solar system or at least have suggested that Mercury and Venus go around the Sun. Similar claims or beliefs are also attributed at around this time to Hicetas of Syracuse (c. 400–c. 335 BC) and Ecphantus (fourth century BC) although it is possible that these were characters in Heracleides' writings, not real people

(continued)

(continued)

Year	Discovery, event, fact, idea, possibility or wild speculation, etc.
c. 350–c. 330 BC	Aristotle (384–322 BC), basing his ideas upon observation of the natural world, advocates the Earth-centred model of the universe. Such is his influence upon subsequent thinking that no alternative is seriously considered by main-stream European workers for nearly two millennia
c. 300 BC	Epicurus (341– 270BC) – *The Extant remains* – *Letter to Herodotus* (Trans. Cyril Bailey) advocates an infinite number of inhabited worlds – "Moreover, there is an infinite number of worlds, some like this world, others unlike it … " and "For nobody can prove that in one sort of world there might not be contained … the seeds out of which animals and plants arise …"
c. 300 BC	Metrodorus of Chios (331–278 BC) – reportedly considered the universe to be infinite and so "A single ear of corn in a large field is as strange as a single world in infinite space." (From Aëtius' *Placita* – which itself is only known from other writings)
c. 270 BC	Aristarchus of Samos (c. 310–230 BC) proposes a Sun-centred model for the solar system, with the planets in their correct order and explains the lack of observed parallax motion for the stars by suggesting that they are so far away that the size of the Earth's orbit is negligible in comparison. He also makes the first good estimate of 60 times the radius of the Earth for the Earth-Moon distance and suggests that the distance to the Sun is 19 times larger than this, although the true value is 400 times larger
c. 100 BC	Lo Hsia Hung advocates a moving Earth as an explanation for the seasons

↑

BC (BCE)

AD (CE)

↓

c. 160	The Syrian writer, Lucian of Samosata, writes the first book to speculate about space travel and alien life. Intended as a satire, his *True History* relates how a group of travellers is carried to the Moon on a giant water spout or whirlwind. The Moon, planets, stars and even the Sun are inhabited by fantastic humanoid type creatures, such as the Vulture Dragoons, Garlic Fighters and Flea Archers. The creatures living on the Moon and Sun are engaged in a war when the travellers arrive
c. 490	Aryabhata (476–550), who was probably born in what is now central India, proposes that the Earth rotates and may have espoused a Sun-centred model for the solar system
c. 800–c. 1200	In *The Adventures of Bulukiya*, a story from *A Thousand and One Nights*, the hero's quest for immortality lead him to travel across space to many other inhabited worlds, some larger than the Earth

(continued)

(continued)

Year	Discovery, event, fact, idea, possibility or wild speculation, etc.
c. 840	Al Farghani (c. 800–c. 870) determines distances to the Sun, Moon and planets. His values, based upon an Earth radius of about 5,000 km are reasonably good for the Moon and Mars (at opposition), but are otherwise much too small
1032	Abū Alī Sīnā (Avicenna – c. 980–1037) observes a dark spot on the Sun which he takes to be Venus in transit although a naked-eye sunspot is also a possibility. If correct, this is the first recorded observation of a transit – the basis of the transit method of detecting exoplanets (Figure 3.2)

FIGURE 3.2   The transit of Venus across the Sun on the 8th June 2004. (Copyright © C. R. Kitchin 2004).

c. 1070	Abū Ishāq Ibrāhīm al-Zarqālī (Arzachel, 1029–1087) notes that Mercury's orbit is oval in shape
1281	Qutb al-Din al Shirazi (1236–1311) publishes his *Nihayat al-idrak fi dirayat al-aflak* in which the planetary orbits are modelled using uniform circular motions

(continued)

(continued)

Year	Discovery, event, fact, idea, possibility or wild speculation, etc.
c. 1350	Ibn al-Shatir (1304–1375) produces a revised version of the Ptolemaic model of the universe that eliminates the need for eccentrics and equants. Although Earth-centred, al-Shatir's system is mathematically equivalent to Copernicus' Sun-centred solar system model
1377	In his *Traité du ciel et du monde* Nicole Oresme queries the prevailing Aristotelian idea of a fixed Earth at the centre of the universe by suggesting that the apparent rotation of the heavens might actually be due to the Earth rotating the other way
1440	Nicholas of Cusa proposes an infinite, centre-less universe, a moving Earth and non-uniform, non-circular motions for the planets in his *De Docta Ignorantia*
1543	*De revolutionibus orbium coelestium* published (though written from 1514 onwards) by Nicolaus Copernicus in which the Sun-centred universe is realistically proposed for the first time
1584	In his *De L'Infinito Universo et Mondi* (Trans. Dorothea Singer), Giordano Bruno extends the Copernican model of the solar system throughout the universe and advocates the idea of an infinite number of solar systems (ideas, which amongst other things, would lead to his being burnt at the stake in 1600 by the Roman Inquisition). For example "Thus is the excellence of God magnified and the greatness of his kingdom made manifest; he is glorified not in one, but in countless suns; not in a single earth, a single world, but in a thousand thousand, I say in an infinity of worlds"
~1570–1601	Tycho (Tyge) Brahe makes large numbers of highly accurate observations of the positions of planets and stars. He used naked eye instruments but by taking great care in their construction and in his observing methods (he conceived the idea of determining and correcting for errors in the instruments for example) was able to achieve positional accuracies of around 3 arc-minutes (Appendix II) – far, far better than any previous work. His data were used by Johannes Kepler to deduce his three laws of planetary motion
1609–1621	Johannes Kepler establishes a firm scientific footing for Copernicus' Sun-centred model of the universe by discovering the three laws of planetary motion. The three laws are: 1. The orbit of a planet is an ellipse with the Sun occupying the position of one of the (two) focuses of the ellipse 2. The planet's speed around its orbits varies in such a way that the line joining the planet and the Sun sweeps out equal areas in equal times (i.e. the planet moves faster when closer to the Sun than when it is further away) 3. The radius of a planet's orbit cubed is proportional to its orbital period squared (strictly the relationship is with half the length of the longest axis of the ellipse – known as the semi-major axis – but the orbits for most solar-system planets are near enough circular for this approximation to be useful) (Figure 3.3)

(continued)

(continued)

Year	Discovery, event, fact, idea, possibility or wild speculation, etc.

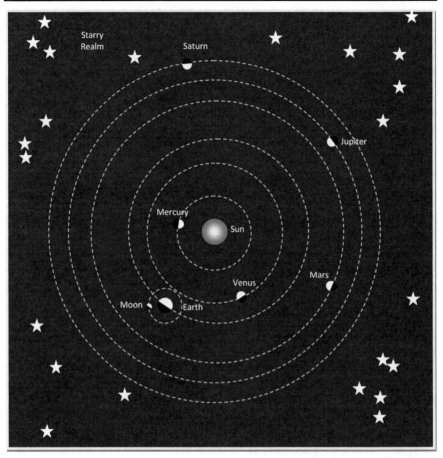

FIGURE 3.3 Kepler's version of Copernicus' heliocentric model of the solar system (not to scale). The introduction by Kepler of elliptical orbits rendered the model realistic for the first time.

Year	Discovery, event, fact, idea, possibility or wild speculation, etc.
1610	Galileo Galilei publishes his *Siderius Nuncius* outlining his early telescopic observations of the heavens. The discoveries announced in this work, such as Jupiter's four satellites, together with later observations, such as of the phases of Venus, convince him of the correctness of Copernicus' Sun-centred model of the universe. Not everyone though, especially the ecclesiastical authorities, concurs, leading to his subsequent trial for heresy
c. 1611	Johannes Kepler uses the idea of space travel to the Moon in his book *Somnium* (not published until 1635). The space traveller, Duracotus, is able to observe the Earth moving through space, thus confirming (in fiction at least) Copernicus' heliocentric solar system model

(continued)

(continued)

Year	Discovery, event, fact, idea, possibility or wild speculation, etc.
c. 1630	Godefroy Wendelin determines the distance of the Sun from the Earth to be about 95 million km – about 60% of the true value
1631	Pierre Gassendi makes the first unequivocal observation of a planetary transit across a stars' disk – the basis for the transit approach to detecting exoplanets. Gassendi observes Mercury as it crosses the face of the Sun following predictions of the event by Kepler
1639	Jeremiah Horrocks makes the first observations of transit of Venus by using a small telescope to project the solar image onto a white sheet of paper
1644	In his *Principia Philosophiae* René Descartes proposes his vortex theory for the motion of the planets
1657	Cyrano de Bergerac writes *L'Autre Monde; ou, les Etats et Empires de la Lune*, followed in 1662 by *les Etats et Empires du soleil* (both published posthumously) describing trips to the Moon and Sun and the societies to be found there
1686	Bernard de Fontenelle's *Conversations on the Plurality of Worlds* speculates that many other worlds (planets) exist and may be inhabited
1687	Isaac Newton's *Philosophiae Naturalis Principia Mathematica* (better known simply as the *Principia*) is published, setting out Newton's law of gravity and laws of motion and deriving Kepler's three laws of planetary motion from the former. The orbital motions of all planets (solar system or exoplanets) are governed by these laws
1698	Christiaan Huygens advocates (posthumously) that many stars have planetary systems in his *Cosmotheoros* – "For then why may not every one of these Stars or Suns have as great a retinue as our Sun, of planets, with their moons, to wait upon them? Nay, there's manifest reason why they should"
1713	Isaac Newton – " ... if the fixed stars are centres of other like systems (i.e. planetary systems), these, being formed by like wise counsel, must all be subject to the domination of One." – *General Scholium* (second edition of the *Principia*)
1728	James Bradley discovers the aberration of starlight – which is the first experimental proof that the Earth is orbiting the Sun
1734	Emanuel Swedenborg proposes in his *Opera Philosophica et Mineralia* that the solar system was formed from a disk of material that had condensed out of the primordial matter – essentially the modern model for the formation of planetary systems (and many other astronomical objects). The model was developed further in 1755 by Immanuel Kant. In 1796, apparently independently, Pierre Laplace proposes the idea again as his nebular hypothesis

(continued)

(continued)

Year	Discovery, event, fact, idea, possibility or wild speculation, etc.
1781	First 'new' planet discovered (i.e. the discovery of a planet not known since pre-historical times). Uranus was discovered through visual telescopic observations by William Herschel (Figure 3.4)

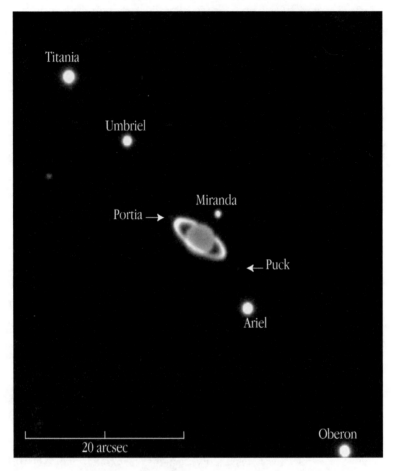

FIGURE 3.4 A near-infrared image of Uranus with 7 of its 27 currently known satellites and its rings obtained using ESO's 8.2-m VLT Antu telescope. (Reproduced by kind permission of ESO).

1782	Algol (β Per – the Greek alphabet is listed in Appendix I for reference) discovered by John Goodricke and Edward Piggott to be an eclipsing binary star from its periodic diminutions in brightness. Although the eclipsing body in this case is another star (but Piggot at least thought that it might be a planet), the discovery foreshadows the transit approach used for detecting exoplanets today

(continued)

Year	Discovery, event, fact, idea, possibility or wild speculation, etc.
1801	Discovery of the first asteroid, Ceres, by Giuseppe Piazzi
1802	William Wollaston observes seven narrow dark regions within the solar spectrum. Wollaston regarded these as the natural boundaries between the colours, but we now call them absorption lines. The lines are produced in all stars' atmospheres by atomic elements and their ions. The change in the lines' positions within a spectrum arising from the star's velocity towards or away from us (the Doppler shift) underlies the method of monitoring stars' radial velocities upon which is based the main method to date of detecting exoplanets (Figure 3.5)

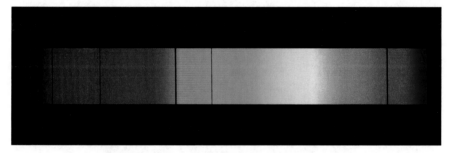

FIGURE 3.5 The solar spectrum as Wollaston might have seen it.

1838	Stellar parallax (the change in position in the sky of a nearby star relative to the positions of very distant stars as the Earth moves around its orbit) is observed for the first time. Freidrich Bessel measures the parallax of 61 Cyg as 0.3136″ – giving its distance as 10.6 light years (just 6% smaller than the modern value). This is the second experimental demonstration (after the aberration of starlight) of the Earth's orbital motion
1842	Christian Doppler publishes his *Über das farbige Licht der Doppelsterne und einiger Gestirne des Himmels* in which he proposes that the different colours sometimes observed for stars in a double or binary system arise from the different velocities of the two stars along the line of sight. The suggestion for the colours of stars was incorrect but the change in wavelength of light towards longer wavelengths when the light emitter moves away from us and to shorter wavelengths when the light emitter moves towards us is real. The effect, now known as the Doppler shift, enables the stars' velocities along the line of sight to be measured and so underlies the radial velocity methods of detecting exoplanets
1844	The first detection of an unseen stellar companion from the orbital movements of the visible object. Freidrich Bessel announces that Sirius has a dark companion. The companion is later found to be another star but this discovery foreshadows the astrometric method of discovering exoplanets (Figure 3.6)

(continued)

(continued)

Year	Discovery, event, fact, idea, possibility or wild speculation, etc.

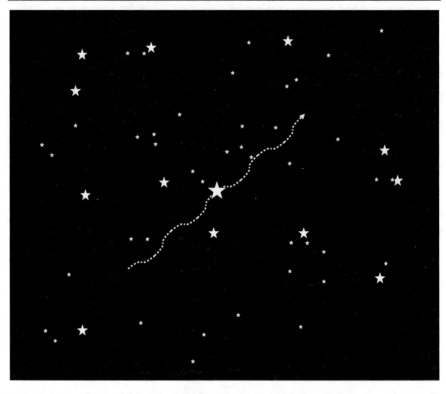

FIGURE 3.6 The wobble in a nearby star's movement with respect to much more distant stars (its proper motion) reveals the presence of an unseen companion whose gravitational pull is shifting the visible star from side to side as the unseen object orbits around it

1846	Discovery of Neptune by Johann Gottfried Galle working at the Berlin observatory and using predictions of its position by Urbain Le Verrier. Le Verrier (and also John Couch Adams) based their predictions upon the observed deviations of Neptune from its predicted orbit, foreshadowing astrometric and transit timing variation ways of detecting and confirming exoplanets
1847	Inter-stellar extinction – now known to be due to dust particles similar to those coagulating to form proto-planets – detected by Freidrich Georg Wilhelm von Struve

(continued)

(continued)

Year	Discovery, event, fact, idea, possibility or wild speculation, etc.
1851	Léon Foucault's pendulum at the Paris observatory is the first experimental proof of the Earth's rotation. The pendulum is free to oscillate in any direction and acts like a gyroscope. It continues to swing in the same direction in space (i.e. relative to the distant galaxies, etc.) while the Earth rotates beneath it. The pendulum constructed by Foucault at the Panthéon in Paris a few weeks later demonstrated the effect to anyone who cared to watch it for a few minutes. This pendulum, which was 67 m in length, had an apparent rotation of 11° per hour (not 15° because of Paris' latitude of 49° – a Foucault pendulum apparently rotates at 15° per hour at the poles and does not rotate with respect to the Earth at the equator)
1855	The first serious, but mistaken, claim for the detection of an exoplanet. William Stephen Jacob of the East India Observatory in Madras thought that apparent orbital anomalies in the binary star 70 Oph might result from an exoplanet
1865	Jules Verne's *De la Terre à la Lune* presages realistic exploration of planets and other solar system objects with some surprisingly good estimates of what will be needed, including the correct calculation of the escape velocity (11 km/s) required to leave the Earth
1868	The first measurement of the radial velocity of a star from its Doppler shift. William Huggins finds the radial velocity of Sirius to be nearly 50 km/s away from us – although the modern value is far smaller and the direction of the motion is towards us (Figure 3.7)
1877	Giovanni Schiaparelli observes dark lines on Mars which he calls channels. Unfortunately the mistranslation of the Italian for channels, *'canali'*, into the English *canals* is taken to imply a completely unintended artificial origin for the features. Schiaparelli's canali are now known to be optical illusions, the affair, though, sparked wholesale but generally ill-informed speculation about Martian and other forms of alien life which continues to the present day
1893	Johannes Wilsing at Potsdam's astrophysical observatory mistakenly claims to have detected a planet orbiting 61 Cygni
1899	Repeated, but again mistaken, claim this time by Thomas Jefferson Jackson See for the detection of a planet belonging to 70 Oph
1902	James Jeans derives the requirements for the gravitational collapse of an inter-stellar gas cloud (thus perhaps leading to the formation of stars and planets). For a given temperature and density of the gas there is a minimum mass (the Jeans' mass) before the cloud can become unstable and start contracting. Perhaps unsurprisingly, the Jeans' mass for the conditions inside a typical inter-stellar gas cloud is about one solar mass

(continued)

(continued)

Year	Discovery, event, fact, idea, possibility or wild speculation, etc.

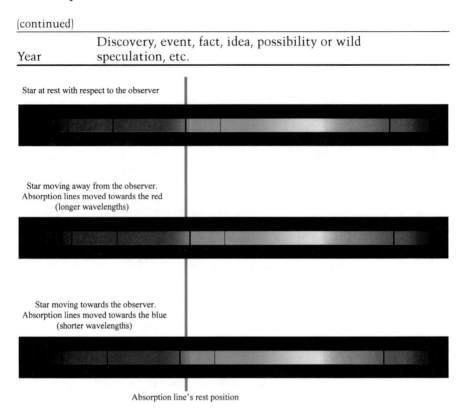

Star at rest with respect to the observer

Star moving away from the observer. Absorption lines moved towards the red (longer wavelengths)

Star moving towards the observer. Absorption lines moved towards the blue (shorter wavelengths)

Absorption line's rest position

FIGURE 3.7 The change in the positions of the absorption lines with respect to the colours (wavelengths) of the spectrum when an object such as a star moves away from or towards the observer. NB. The changes in the positions of the lines are much exaggerated.

Year	
1903	Svante Arrhenius presents the first detailed scientific case for life originating from outer space. The theory, called Panspermia, envisages spores pervading the whole of space and being moved between planets by radiation pressure from the stars. On arrival at a suitable site the spores spring into life. More spores would be produced when the seeded planets developed dense populations of life forms and some of these were lost to space having been carried to the top of the planets' atmospheres by convection currents etc. The origin of the 'first spore' is unexplained. A restricted version of the theory, called exogenesis, postulates the origin of life on Earth via transfer from elsewhere in the universe, but does not speculate as to whether this is a common or a rare process. The basic concepts of panspermia had earlier been mooted by, amongst others, Anaxagoras, Kelvin and Helmholtz

(continued)

(continued)

Year	Discovery, event, fact, idea, possibility or wild speculation, etc.
1904	Johannes Hartmann detects the presence of inter-stellar matter through the absorption lines due to calcium that it produces in the spectrum of δ Ori. The dense concentrations of inter-stellar matter known as giant molecular clouds are the birth places of stars and planets
1905	Forest Ray Moulton and Thomas Chrowder Chamberlin propose that the planets formed from solar material ejected from the Sun by tidal forces from another star combined with solar prominence activity. Although later discarded, the theory also included the idea of the condensation of planetesimals and their coalescence into larger objects which is still a part of the current theory of planet formation
1919	James Jeans proposes in his *Problems of Cosmogony and Stellar Dynamics* that the planets were formed from a strand of material tidally wrenched from the proto-Sun by a close encounter with another star. Since close passages between two stars are extremely rare this would make the existence of exoplanetary systems extremely rare as well. Subsequent work however has shown that Jeans' proposal is most unlikely to lead to the formation of planets and so contraction from a nebulosity is the currently favoured process (Figure 3.8)
1930	Discovery of Pluto by Clyde Tombaugh. For a long time classed as the ninth planet of the solar system, Pluto is now designated as a dwarf planet and is probably just one of the larger members of the Kuiper belt objects (KBOs)
1930	Bernard Lyot builds the first coronagraph. Lyot's instrument is designed to observe the faint solar corona adjacent to the much brighter Sun – adaptations of the design are now used to enable the direct imaging of faint exoplanets adjacent to their much brighter stars
1942	A mistaken claim, by Kaj Strand, for an exoplanet orbiting 61 Cyg A
1943	Yet another mistaken claim, this time by Dirk Reuyl and Erik Holberg, for the detection of a planet with ten times the mass of Jupiter belonging to 70 Oph
1944	Carl Friedrich von Weizsäcker revises Kant's nebular hypothesis of the formation of planetary systems to more-or-less its present form. The modern nebular hypothesis, while not without its problems still, is now generally accepted as the probable way in which planets are created
1947	Bart Bok and Edith Reilly discover the small dense clouds of gas and dust now known as Bok Globules. Bok globules are usually embedded inside H II regions and in many cases are in the process of collapsing to form proto stars and planets (Figure 3.9)

(continued)

(continued)

Year	Discovery, event, fact, idea, possibility or wild speculation, etc.

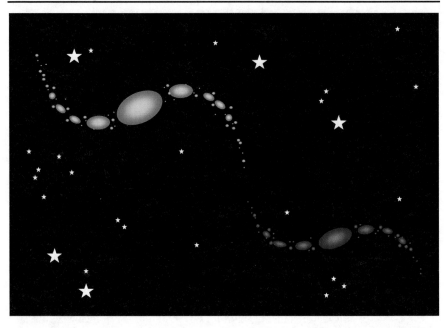

FIGURE 3.8 Jeans', now discredited, suggestion that planetary systems form from the material dragged out from two stars passing close by each other. In this schematic illustration the two stars have passed each other and have become elongated by the effects of gravity. Like the tides raised in the Earth's oceans by the Moon, there are two high tides in each star; one near the point on the star's surface that is closest to the other star and the second on the star's opposite side. Material is thus wrenched from the stars from both of these points and may go into orbit around the stars or be lost to interstellar space.

Year	Discovery, event, fact, idea, possibility or wild speculation, etc.
1950	Enrico Fermi points out that at a speed of 0.1% that of light (300 km/s) an alien society could spread throughout the entire Milky Way galaxy in just a 100 million years. The Fermi paradox poses the question why therefore, given that the Milky Way is at least a hundred times older than this time interval, do we see no sign of any such alien activity? One answer suggests that we are the first intelligent life forms to evolve in the galaxy, another that intelligent life self-destructs through over use of resources in a short time, another that aliens are indeed out there, but that they are hiding or that we are not looking for them in the right way
1951	A mistaken claim, by Peter (Piet) van de Kamp and Sarah Lippincott, for an exoplanet orbiting Lalande 21185

(continued)

(continued)

Year	Discovery, event, fact, idea, possibility or wild speculation, etc.

FIGURE 3.9 The Bok Globule Barnard 68 (B68). The image was obtained using the 8.2-m Antu telescope of ESO's VLT. The cloud is about the same size as the Oort cloud of the solar system (0.2 light years) and has a mass about twice that of the Sun. (Reproduced by kind permission of ESO).

1952	Otto Struve proposes the radial velocity (Doppler) method for detecting exoplanets. Currently about three-quarters of the known exoplanets have first been found through the use of this approach
1957	Kaj Strand repeats his claim of an exoplanet orbiting 61 Cyg A and gives a mass eight times that of Jupiter, an orbital radius of 2.4 astronomical units and a period of 4.8 years. Modern observations fail to confirm this and place an upper limit of 2.1 Jupiter masses on any exoplanet out to 5 astronomical units from the star
1960	Frank Drake attempts to predict the number of extraterrestrial civilisations with which we might potentially be able to come into contact. This first version of the 'Drake Equation' suggests an answer of two or three such civilisations

(continued)

(continued)

Year	Discovery, event, fact, idea, possibility or wild speculation, etc.
1960	Frank Drake uses the Green Bank 26-m radio telescope to search for signs of life at τ Cet and ε Eri. Project Ozma was the first SETI experiment and scanned a 400 kHz waveband close to the 21-cm line. No evidence of artificial signals from near either of the two stars was detected after 400 h of observations
1960	Freeman Dyson argues that some, perhaps most, intelligent forms of alien life will eventually need to utilise all the energy available from their host stars. He suggests that they could do this by deconstructing their planetary system and building from the material a solid spherical shell (the Dyson sphere) around the star. However in the absence of anti-gravity it would be impossible to construct and would have the probably undesirable side-effects of the inhabitants falling into the central star and of the sphere colliding with the star in a short space of time
	Later, Larry Niven in his *Ringworld* novels suggests modifying the sphere into a gigantic ring that would be spun to provide artificial gravity. Although more feasible in some ways than the Dyson sphere, the ring would need to be continuously driven in some manner to keep it centred on the star. Furthermore the construction material would need to be around a hundred thousand times stronger than multi-walled carbon nanotubes (the strongest material that we currently know about) if the ring were to rotate rapidly enough to provide the equivalent of the Earth's gravity (Figure 3.10)
	A just-about-possible variation of the Dyson sphere called the Dyson swarm, involves parcelling out the matter of the planetary system into a myriad of small bodies in independent (and non-intersecting) orbits. All the star's radiation emitted in any direction would then be intercepted by one or another of these bodies. Such an object might be observable in the infrared when the intercepted radiation is re-emitted from the dark sides of the small bodies. For the solar system and for human-comfortable temperatures, the concept would imply some 100,000,000,000,000,000,000,000 (10^{23}) 2-m diameter bodies with a mean orbital radius of 1.7 astronomical units if all its planetary matter could be so used. The mean temperature would be around 20°C and it would radiate on the outside in the medium infrared, peaking at a wavelength of about 10 μm
1960	Another mistaken claim, by Peter van de Kamp and Sarah Lippincott, for an exoplanet orbiting Lalande 21185
1963	Peter van de Kamp mistakenly claims to have discovered an exoplanet around Barnard's star
1966	Captain James T. Kirk – "The mission of this vessel is to explore strange new worlds; to seek out new life and new civilizations." The debut of the iconic TV series 'Star Trek' results in enormous public interest in and the widespread acceptance of the likely existence of exoplanets and of life forms upon them

(continued)

(continued)

Year	Discovery, event, fact, idea, possibility or wild speculation, etc.

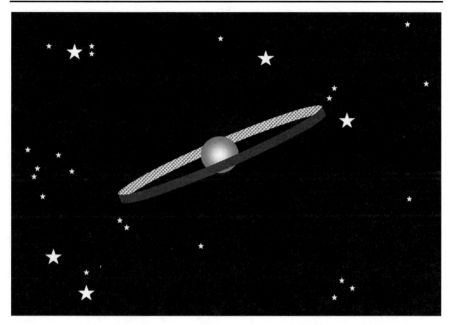

FIGURE 3.10 Niven's *lebensraum* solution – the Ringworld concept.

1967	Jocelyn Bell et al. discover regular pulses at radio wavelengths coming from a source in Vulpecula. Originally, and perhaps not entirely seriously, attributed to 'Little Green Men', this source and other similar ones are quickly identified as pulsars and not the first signal from alien life (Figure 3.11)

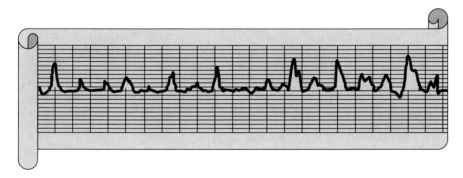

FIGURE 3.11 The first signal to us from Little Green Men? – No – It is the radio signal from a source in Vulpecula that was originally seen by Jocelyn Bell on a chart recording and which revealed regular pulses at one and a third second intervals – but it turned out to be the pulsar CP1919.

(continued)

(continued)

Year	Discovery, event, fact, idea, possibility or wild speculation, etc.
1969	Mistaken claim by David Richards et al. to have found an exoplanet orbiting the pulsar in the Crab Nebula (CP0532)
1973	Using a sophisticated method to analyse earlier data Oliver Jensen and Tadeusz Ulrych mistakenly claim to have detected five exoplanets orbiting Barnard's star
1974	Gösta Gahm et al. attribute light variations in the T Tauri star, RU Lup, to dust concentrations within the star's surrounding envelope of gas and dust. Such concentrations could be the precursors of exoplanets, but later work suggests that the concentrations would be unstable
1975	Nick Scoville et al. discover the Sgr B2 giant molecular cloud. GMCs are the main sites of star and planet formation
1977	A mistaken claim by Alexander Deutsch et al. for the detection of two exoplanets orbiting 61 Cyg A and one orbiting 61 Cyg B, with masses 6, 12 and 7 times that of Jupiter. Modern observations rule out any planets around the two stars out to 5 astronomical units with masses greater than 2.1 Jupiter masses (61 Cyg A) and 1.5 Jupiter masses (61 Cyg B)
1979	Marek Demianski and M. Proszynski suggest that an exoplanet is one possible explanation for the 3-year periodicity in the timings of pulsar PSR 0329+54. Later work does not confirm the suggestion
1979	Bruce Campbell and Gordon Walker devise a spectrograph that uses the absorption lines of hydrogen fluoride to enable the radial velocity of a star to be measured to an accuracy of ± 15 m/s. Using this instrument they start searching for exoplanets via the radial velocity method
1983	Infrared detection by the IRAS spacecraft of circumstellar disks around β Pic, Vega, Fomalhaut and ε Eri (Figure 3.12)
1984	Bradford Smith and Richard Terrile obtain direct visible-light images of the circumstellar disk around β Pic using the du Pont 2.5-m telescope
1984	Mistaken claim by Donald McCarthy et al. for the discovery of a brown dwarf orbiting VB8 in Ophiuchus
1987	Ben Zuckerman and Eric Becklin detect excess infrared radiation from the white dwarf Giclas 29–38 which they attribute to the presence of a brown dwarf. Later work reveals that the excess emission is from a cloud of dust
1988	The first detection of an exoplanet, though not confirmed until 2003. Based upon 6 years of observations of the star's radial velocity variations, Bruce Campbell, Gordon Walker and Stephenson Yang claim to have detected an exoplanet orbiting γ Cep A with a period of about 2.7 years. The claim however was withdrawn in 1992. Then in 2003, based upon 20 years of accumulated data the existence of an exoplanet with an orbital period of 2.5 years and a mass of at least 1.7 Jupiter masses was confirmed. The same team also tentatively identified exoplanets around ε Eri (1992) and β Gem (1993) but these were also unconfirmed until much later

(continued)

(continued)

Year	Discovery, event, fact, idea, possibility or wild speculation, etc.

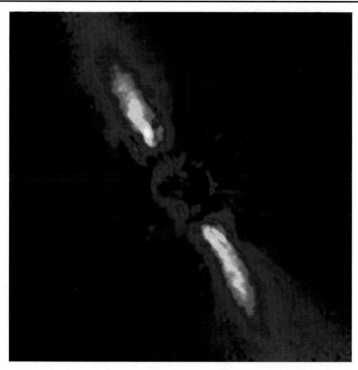

FIGURE 3.12 Circumstellar disk around β Pic. The star has been obscured by an occulting disk so that the much fainter disk shows clearly in this near infrared image obtained using ESO's 3.6-m telescope. The disk extends out to at least 1,800 astronomical units from the star. (Reproduced by kind permission of ESO).

Year	
1989	Discovery of a possible exoplanet for the star HD 114762 by David Latham from the star's radial velocity variations. However the object's mass could range from 11 to 145 Jupiter masses depending on the inclination of its orbit to the line of sight so it could be a brown dwarf or even a red dwarf star instead of an exoplanet. If, as seems most likely, it is confirmed to be a brown dwarf, then it will be the first of this class of objects to have been discovered
1991	Andrew G. Lyne mistakenly claims to have discovered an exoplanet around the pulsar PSR 1829–10
1991	Shude Mao and Bohdan Paczyński propose the gravitational microlensing method of detecting exoplanets
1992	First definitive detection of exoplanets. Aleksander Wolszczan and Dale Frail discover two rocky planets orbiting the pulsar PSR B1257+12 from the changes in the arrival times of the radio pulses from the pulsar

(continued)

(continued)

Year	Discovery, event, fact, idea, possibility or wild speculation, etc.
1992	David Jewitt and Jane Luu discover the first object in a planetary-type orbit further out from the Sun than Pluto. Given the designation, (15760) 1992 QB1, it is about 160 km across and takes 290 years to complete a single circuit of its 44 astronomical unit orbit. It is now classed, along with Pluto, Charon and about another thousand similar entities detected since 1992, as a Trans-Neptunian Object (TNO)
1993	Stephen Thorsett et al. suggest that an exoplanet might be orbiting the pulsar/white dwarf binary PSR B1620-26 based upon variations in the timing of the pulsar's pulses. A third star within the system, however, would have, at that time, explained the data equally well. The status of the object as an exoplanet was not confirmed until 2003
1994	Tadashi Nakajima et al. make the first discovery of a brown dwarf using the 1.5-m telescope at Mount Palomar. The brown dwarf may be seen directly on the confirmatory red-light image obtained using the HST in November 1995. It has a mass 20–50 Jupiter masses and orbits a red dwarf star, Gliese 229, which is about 16 light years away from us in Lepus (Figure 3.13). Before this brown dwarf had been confirmed, Rafaele Rebolo et al. announced in September 1995 from observations made using the 0.82-m telescope at the Instituto de Astrofisica de Canarias (IAC) that Teide 1, some 400 light years away from us in the Pleiades, had been verified to be a brown dwarf with a mass of 55 Jupiter masses
1995	First definitive detection of an exoplanet belonging to a 'normal' star. From changes to the star's radial velocity Michel Mayor and Didier Queloz detect an exoplanet with about half the mass of Jupiter orbiting the main sequence star 51 Peg
1996	David McKay et al. announce the discovery of biosignatures, including possible bacterial remains, within meteorite ALH 84001. The meteorite is thought to have come from Mars, being blasted off the planet during a large meteorite impact. Subsequent work has failed to confirm the claim, with all of the features explainable by inorganic processes
1998	Discovery by Motohide Tamura et al. using the Subaru telescope of planetary-mass objects that are unassociated with a more massive brown dwarf or star. Many such objects are now known and are variously called sub-brown dwarfs, free-floating planets or planetars
1998	Susan Terebey et al. claim the first direct image of an exoplanet. The object, TMR-1C was imaged by the HST and appears linked by a streamer of material to a binary star suggesting the possibility that it has been ejected from that system by gravitational perturbations. Later the claim is withdrawn. However in 2009 further observations with the Canada-France-Hawaii telescope (CFHT) re-open the discussion by suggesting that TMC-1C could be a young proto-planet embedded in a disk of dusty material

(continued)

(continued)

Year	Discovery, event, fact, idea, possibility or wild speculation, etc.

FIGURE 3.13 The confirmatory HST direct image of the brown dwarf, Gliese 229B. The main star is off to the left, but its glare is still apparent. The brown dwarf is the small dot near the centre. (Reproduced by kind permission of S. Kulkarni (Caltech), D. Golimowski (JHU) and NASA).

Year	Discovery, event, fact, idea, possibility or wild speculation, etc.
1999	First detection of an exoplanet via its star's brightness variations as the planet crosses in front of (transits) the face of the star (HD 209458).The exoplanet was discovered via the radial velocity method
1999	First direct detection of an exoplanet's spectrum. Andrew Collier-Cameron et al. used the 4.2-m William Herschel telescope to observe the spectrum reflected from the eight Jupiter mass exoplanet orbiting τ Boö Doppler shifted by 75 km/s from the star's spectrum
2001	Sodium detected in the atmosphere of the exoplanet orbiting HD 209458 by David Charbonneau et al. using the HST's spectroscope
2001	Nuno Santos et al. using the Coralie spectroscope detect the first exoplanet to be found within its star's habitable zone. The planet though (HD28185b), has a mass of six Jupiter masses

(continued)

(continued)

Year	Discovery, event, fact, idea, possibility or wild speculation, etc.
2002	Discovery of the first 'cold Jupiter' exoplanet around 55 Cnc by Geoff Marcy and Paul Butler. The planet has a similar mass and distance from its star as does Jupiter from the Sun
2002	First discovery of exoplanets via the transit method. OGLE-TR-10-b and OGLE-TR-56-b were identified as likely exoplanets by Andrzej Udalski et al. Confirming radial velocity observations were made over the next 2 years
2003	Discovery of Eris (announced in 2005) by Mike Brown et al. using the Samuel Oschin Schmidt telescope, the object is the largest dwarf planet known within the solar system and a member of the Kuiper belt
2003	Discovery of the 100th exoplanet
2003	Confirmation of the reality of the first exoplanet to be detected (in 1988 orbiting the star γ Cep A)
2003	Kevin Volk et al. discover a second brown dwarf in the ε Indi system. The two brown dwarfs have masses around 30 and 50 Jupiter masses and orbit each other with a separation of just 2 astronomical units – making this the first known brown dwarf binary system. The brown dwarfs orbit the ε Indi star at a distance of 1,500 astronomical units
2004	First undisputed direct image of an exoplanet. The planet, with a mass five times that of Jupiter orbits 55 astronomical units out from the brown dwarf 2M1207 in Centaurus. The near-infrared images were obtained using the VLT. Confirmation that it was an exoplanet, and not a background object, came in 2005 from observations of its movement via VLT and HST images (Figure 3.14)
2004	First discovery of an exoplanet via the gravitational microlensing technique. Ian Bon et al. use the MOA 1.8-m telescope and the OGLE camera to detect a 2.6 Jupiter mass planet orbiting a star 17,000 light years away from us in Sagittarius
2005	The first detection of direct radiation from exoplanets orbiting normal stars. The difference between the combined infrared intensities of the stars and their planets together compared with that of the stars alone when the planets are behind the stars (secondary eclipse) is detected using the Spitzer space telescope for the exoplanets, HD 209458b and TrES-1
2005	Theodor Hänsch and John Hall part share the Nobel Prize for physics for their development of lasers able to emit their light in the form of pulses with very short durations. These lasers form the basis of the laser comb comparison spectrum used in the radial velocity approach to detecting exoplanets
2006	Discovery of the 200th exoplanet
2006	Launch of the CoRoT (Convection, Rotation and planetary Transits) spacecraft. It mission is to detect new exoplanets via the transit method as well as studying planetary interiors via astroseismology

(continued)

(continued)

Year	Discovery, event, fact, idea, possibility or wild speculation, etc.

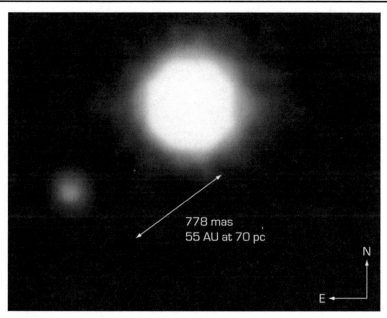

778 mas
55 AU at 70 pc

N

E

FIGURE 3.14 An infrared image of 2 M1207 obtained using the 8.2-m Yepun telescope of ESO's VLT showing the first direct image of an exoplanet – the planet is the fainter object left of centre. (Reproduced by kind permission of ESO).

Year	Discovery, event, fact, idea, possibility or wild speculation, etc.
2007	Water vapour detected by the Spitzer space telescope in the spectrum of the exoplanet HD 189733b
2007	First exoplanet, CoRot-1b, discovered by the CoRoT spacecraft. The planet is a hot Jupiter orbiting the G-type star in Monoceros, CoRoT-1
2007	Heather Knutson et al. Produce the first map of an exoplanet. The hot Jupiter orbiting the red dwarf HD 189733 in Vulpecula was observed for 33 h – more than half its orbital period – using the Spitzer spacecraft. The map has a resolution of about a quarter of the planet's radius (20,000 km) and shows a hot spot off-set by about 30° of longitude from the sub-stellar point. The planet's temperature ranges from 970 to 1200 K and easterly winds moving at speeds of nearly 10,000 km/h may be needed to ensure such a temperature distribution

(continued)

(continued)

Year	Discovery, event, fact, idea, possibility or wild speculation, etc.
2008	First visible-light direct image of a normal star's exoplanet. In fact six such exoplanets were announced between September and November – four of them confirmed and two probables (Figure 7.3). Direct HST visible light images of a confirmed 0.5 to 2 Jupiter mass exoplanet were obtained by Paul Kalas and James Graham for the star Fomalhaut. Simultaneously *three* confirmed Jupiter-mass exoplanets were announced for the star HR 8799 by Christian Marois and his team who had observed it in the infrared with both the 10-m Keck and 8.1-m Gemini telescopes. Other teams at about the same time imaged in the infrared an eight Jupiter mass planet 300 astronomical units out from 1RXS 1609 (confirmed in 2010) and an eight Jupiter mass exoplanet in a ~17 to ~35 year orbit 8–15 astronomical units out from β Pic (also confirmed in 2010)
2008	The first identification of an organic molecule (methane) on an exoplanet from observations made by Mark Swain et al. of the spectrum of HD189733b using the HST
2008	Discovery of the 300th exoplanet
2008	Svetlana Berdyugina et al. make the first detection of polarized light scattered within an exoplanetary atmosphere. The polarisation of the hot Jupiter orbiting HD 189733 showed maxima when the planet was at its greatest angular separations from the star
2009	Launch of the Kepler spacecraft whose mission is to detect Earth-like exoplanets via transits (Figure 3.15)

FIGURE 3.15 An artist's impression of the Kepler spacecraft silhouetted against an imaginary exoplanetary system. (Reproduced by kind permission of NASA, the Kepler mission and Wendy Stenzel).

(continued)

(continued)

Year	Discovery, event, fact, idea, possibility or wild speculation, etc.
2009	Discovery of the 400th exoplanet
2009	Another refuted detection of an exoplanet by astrometry. After 12 years of observations using the Hale 5-m telescope, Steven Pravdo and Stuart Shaklan suggested that the minute changes in the position in the sky of an M-type dwarf star, VB10 in Aquila, were due to a six Jupiter mass planet in orbit around it. Subsequent radial velocity measurements ruled out the possibility of a planet with a mass greater than three Jupiter masses
2009	The first exoplanet discovered in a retrograde (i.e. moving in the opposite direction to the rotation of its host star) orbit. The planet, WASP-17b was detected by Coel Hellier and his team using the SuperWASP South instrument and confirmed via Doppler shift measurements made with the Coralie and HARPS spectrographs
2009	Anthony Colaprete et al. reveal that the LCROSS lunar mission has finally confirmed the presence of water on the Moon, thus potentially making possible the establishment of a long-term or permanent manned outpost on the Moon
2009	Jamie Elsila et al. announce that glycine, the simplest amino acid and one of the building blocks of proteins and DNA, has been discovered in samples recovered from Comet Wild 2 by the Stardust spacecraft. The discovery lends support to the possibility of life originating in space
2010	First discoveries of exoplanets by the Kepler spacecraft (five hot Jupiters)
2010	First spectrum of an exoplanet imaged separately from that of its star. Marcus Janson et al. used the VLT to pick up the spectrum of the middle exoplanet of the three known to orbit the hot solar-type star in Pegasus, HR 8799
2010	Ben Burningham and Sandy Leggett find the lowest temperature brown dwarf to date. SSDS1416+13B has an estimated surface temperature of about 500 K – far cooler than many hot-Jupiter type planets (Figure 3.16)
2010	First pair of direct images showing an exoplanet moving from one side of its host star to the other. β Pic b was imaged in 2003 and 2009 by ESO's VLT on either side of its star (Figure 7.3d)
2010	The probable exoplanet for 1RXS 1609 confirmed
2010	First direct spectroscopic measurement of the orbital velocity of an exoplanet (HD 209458 b) and the first direct measurement of the speed of the winds in its atmosphere. Ignas Snellen et al. used ESO's VLT to study the planet's carbon monoxide lines during a transit. The gas in the atmosphere is streaming at between 5,000 and 10,000 km/h from the sub-stellar point around to the dark side

(continued)

(continued)

Year	Discovery, event, fact, idea, possibility or wild speculation, etc.

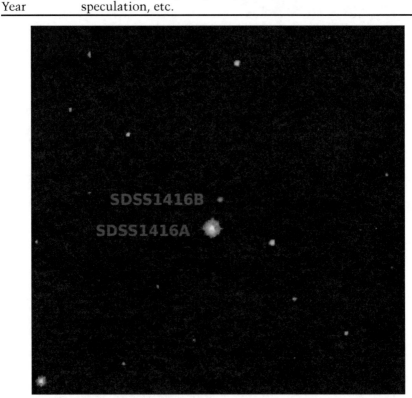

FIGURE 3.16 A UKIRT image of the brown dwarf binary SSDS1416+13 A and B. The latter has the lowest recorded temperature for any known brown dwarf – just over 200°C. (Reproduced by kind permission of JAC/ UKIRT, University of Hertfordshire).

Year	
2010	Christophe Lovis et al. detect an exoplanetary system with up to seven planets using the HARPS spectrograph on ESO's 3.6-m telescope. Six Neptune-sized planets belonging to the solar-twin star, HD 10180, in Hydra (Figure 3.17) have been confirmed. The planets' orbital periods range from 6 days to 2 years. Another planet, perhaps just 40% more massive than the Earth, still awaits confirmation
2010	The first exoplanetary system discovered by transits. The Kepler spacecraft detects two transiting Saturn-mass exoplanets around the solar-type star, Kepler-9, in Lyra. A third exoplanet in the system – a possible super-Earth – is confirmed later in 2010
2010	Based upon a Keck survey of 166 stars, Andrew Howard and Geoff Marcy predict that 23% of solar-type stars will possess exoplanets with similar masses to that of the Earth

(continued)

(continued)

Year	Discovery, event, fact, idea, possibility or wild speculation, etc.

FIGURE 3.17 An artist's impression of the exoplanetary system belonging to HD 10180 viewed from near the Neptune-mass third exoplanet out from the host star, HD 10180d. (Reproduced by kind permission of ESO and L. Calçada).

Year	Discovery, event, fact, idea, possibility or wild speculation, etc.
2010	Red dwarf stars shown to be up to 20 times more abundant relative to solar type stars in elliptical galaxies than they are within the Milky Way. Pieter van Dokkum et al. observed eight elliptical galaxies using the Keck telescopes and concluded "There are possibly trillions of [exo] Earths orbiting these stars"
2010	The first analysis of the atmosphere of a super Earth. Jacob Bean et al. observed GJ 1214 b using the VLT during a transit. The near infrared spectrum of the atmosphere turned out to be featureless, ruling out hydrogen as a primary component of the atmosphere. The researchers suggest that the atmosphere either has a thick high level cloud layer that masks any hydrogen that may be present or that it contains a high proportion of water vapour (steam)

(continued)

(continued)

Year	Discovery, event, fact, idea, possibility or wild speculation, etc.
2010	Discovery of the first exoplanet from another galaxy. The planet, HIP 13044 b, is actually now a part of the Milky Way but when it was born it belonged to a dwarf galaxy once orbiting our galaxy but now ripped apart by tides. The planet's host star is a part of the Helmi Stream, an association of 10–100 million old stars that loops several times around the Milky Way. The exoplanet is a hot Jupiter that just escaped being engulfed by its host star when the latter became a red giant but which may suffer that fate in a few million years when the star expands again
2010	Discovery of the 500th exoplanet
2011	Second data release from the Kepler spacecraft which includes the confirmed discovery of a six-exoplanet system (Kepler-11 b to g) and over 1,200 possible other exoplanets whose existence still awaits confirmation. 54 of the unconfirmed exoplanets lie within their stars' habitable zones and one of these, KOI 326.01, is probably similar to the Earth in size or smaller
2012	Expected launch of Gaia – ESA's spacecraft that amongst other aims is intended to discover numerous new exoplanets
2012	ALMA (Atacama Large Millimeter Array) expected to be fully operational with the potential to detect large exoplanets directly
2012	ESO's SPHERE instrument for the VLT expected to start operating with the potential for the direct imaging of tens of giant exoplanets. The GPI instrument for the Gemini South telescope is also expected to start operations this year
2013–2015	Discovery of the 1,000th exoplanet to be expected (Figure 3.18)
2014	Expected launch of the JWST – it should be capable of observing exoplanet spectra as well as detecting them via the transit method
2017–2018	Possible launch of ESA's SPICA spacecraft
2018	Expected completion dates for the Giant Magellan Telescope (7 × 8.4-m mirrors) and the European Extremely Large Telescope (42-m diameter mirror)
2018–2025	Possible launch of NASA's SIM Lite spacecraft
2025	Expected completion date for the Thirty-Meter Telescope
2020–2045	Discovery of the 10,000th exoplanet to be expected
2020–2045	Discovery of the first true Earth-twin exoplanet to be expected
2024	The Square Kilometer Array (SKA) expected to become fully operational with the ability to detect exoplanets directly at long wavelengths
2030–2100	Discovery of the 100,000th exoplanet to be expected
2040–2250	Discovery of the 1,000,000th exoplanet to be expected

(continued)

(continued)

Year	Discovery, event, fact, idea, possibility or wild speculation, etc.

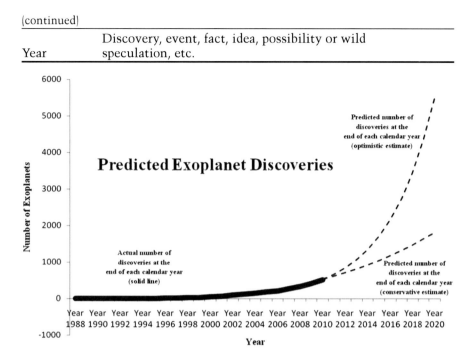

FIGURE 3.18 A graph showing the increase in the number of known exoplanets with time (data from – The Extrasolar Planets Encyclopaedia – http://exoplanet.eu/). Two possible extrapolations are shown. The 'optimistic' line suggests that the 1,000th exoplanet will be found in early 2013. The more conservative estimate suggests that this will take until 2015.

4. In the Beginning – The First Exoplanet Discoveries

Introduction

Detecting exoplanets is very difficult, which is why success has only been achieved within the last couple of decades or so. An exoplanet is small, of low mass and very faint compared with its host star and it is right next door to that host star. So if you try to look for the exoplanet directly, its light will be swamped by that emitted by the star. If you try to detect the star changing its position in the sky as the exoplanet moves around it, then that movement will be buried in the uncertainties in your measurements of the star's position. Likewise, if you try to detect the star's changing velocity as the exoplanet orbits its host star, then those changes will also be buried in the uncertainties in your measurements of the star's spectrum.

Successful detection of exoplanets had to await the development of techniques and instruments that reduced the uncertainties in the measurements to unbelievably low levels – and even now exoplanet detection remains a struggle. Nonetheless exoplanets *are* being found – and by several different methods (Figure 4.1). The most successful approach by far is called the Radial Velocity or Doppler Method – and about three-quarters of the exoplanets that we currently know about have been found using it.

We usually think of planets orbiting the star – like the Earth going around the Sun. But this is actually an incorrect picture. Just as the star pulls the planet towards itself via gravity, so the planet pulls the star towards itself. The result is that both the planet and the star are moving in orbits around a point that we call the centre of gravity (or centre of mass or barycentre).

C. Kitchin, *Exoplanets: Finding, Exploring, and Understanding Alien Worlds*, Astronomers' Universe, DOI 10.1007/978-1-4614-0644-0_4,
© Springer Science+Business Media, LLC 2012

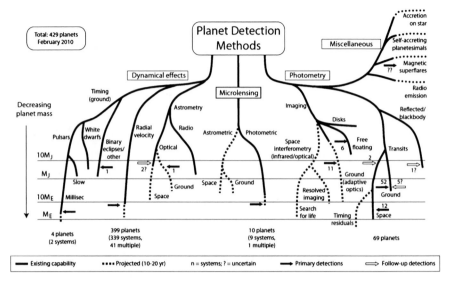

FIGURE 4.1 A summary of the various methods for detecting exoplanets – both those currently in use and potential future approaches, with their likely limits of detection and current success levels. (Reproduced by kind permission of M. Perryman).

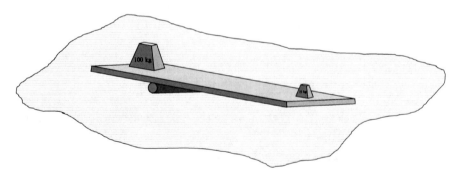

FIGURE 4.2 When the beam is in balance, the roller underneath it is at the centre of gravity between the two masses. The centre of gravity is much closer to the larger mass.

Since the star is so much more massive than the planet, the centre of gravity is much closer to the star than it is to the planet (Figures 4.2 and 4.3). In the case of the Earth and Sun, the centre of gravity is just 500 km away from the centre of the Sun, and 150,000,000 km from the centre of the Earth. Thus as the Earth goes around its orbit, so the Sun moves in a very slight ellipse about 1,000 km across, always on the opposite side of the centre of gravity

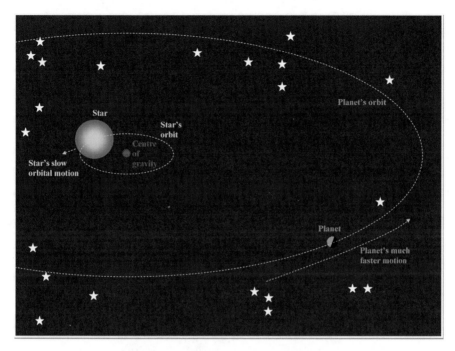

FIGURE 4.3 A star and a planet in their mutual orbits around their common centre of gravity.

from the Earth (Figure 4.3). In fact the Sun's motion is much more complicated than this, because it is simultaneously in smaller or larger orbits for all of the planets, dwarf planets, asteroids, etc.

Even for the solar system's largest planet, Jupiter, the centre of gravity with the Sun is only just outside the surface of the Sun – 780,000 km out from the Sun's centre. Nonetheless, a distant alien astronomer with astronomical equipment comparable with our own might be able to detect the Sun's velocity changing by ±13 m/s as Jupiter moves around its orbit. The alien astronomer might also be able to detect the velocity changes arising from Saturn, Uranus, Neptune and even the Earth and Venus and so detect their presences. For the small solar system objects, the Sun's movement is too small to measure even from our position here on Earth.

How then might the alien astronomer detect this changing velocity of the Sun? Unless our alien has developed a radically different science from our own, it will be by the same method that human astronomers have used since 1868. In that year, Sir William Huggins (Box 4.1) made the first measurement of the

Box 4.1 The First Astrophysicist – Sir William Huggins (1824–1910)

A man of considerable wealth, Huggins (Figure 4.4) was able to devote himself full-time to his interests in many aspects of science, finally in his early 30s deciding to specialize in astronomy. He built his own observatory in Tulse Hill, London, where aided by his neighbour William Miller and later by his wife Margaret, he started observing the spectra of celestial objects. Amongst many signal discoveries, he showed that the stars were composed of the same elements that we find on the Earth, but in the form of very hot gases, that some of the nebulous objects in the sky were gas clouds, others collections of stars, and measured the radial velocities of many stars from the Doppler shifts of the lines in their spectra.

FIGURE 4.4 Sir William Huggins (Proc. Roy. Soc. 86 A, 1911. Reproduced by kind permission of Lafayette photography, Dublin and Cambridge).

velocity of a star (Sirius). He utilized the fact, first proposed by Christian Doppler in 1842, that if a star moves away from us, the absorption (dark) lines in its spectrum are moved to longer wavelengths (and to shorter wavelengths if it moves towards us). Huggins compared the position of a hydrogen absorption line in Sirius' spectrum with that of the same line produced in emission (bright) in his observatory by a Geissler tube (an early type of emission lamp). He found that the star's line was positioned 0.08 nm further towards the red (longer wavelength) end of the spectrum than the artificial line. This measurement implied that Sirius was moving away from us at just less than 50 km/s. However in March, when Huggins was making his observations, the Earth's orbital motion is taking it away from Sirius at about 25 km/s. So Huggins' measurement gave Sirius as moving away from the solar system as a whole at about 25 km/s. Modern measurements, in fact, give Sirius as approaching us at 7.6 km/s, so Huggins measurement was a bit inaccurate. Nonetheless Huggins' method is still the one used today to determine the radial velocities of the stars and has now been developed to the point where it can detect the orbital motions of stars arising from their planets.

The study of the spectra of stars (and those of galaxies, nebulae, planets, comets, etc.) provides the modern astrophysicist with a huge wealth of data on temperatures, compositions, sizes, masses, structures, rotations and many other features of the stars as well as their radial velocities. The study of spectra is the science called spectroscopy and the instruments used to obtain the spectra are called spectrographs or spectroscopes. Since Huggins' early simple visual spectroscope the design of the instruments has advanced enormously and the spectrographs built for large modern telescopes often weigh many tons and cost a good fraction of the price of the telescope (Figure 4.5).

Most people have seen a rainbow and many will know that a glass prism splits white light up into its component colours. Today though, almost all astronomical spectrographs use a diffraction grating to produce the spectrum. The diffraction grating is a mirror which has had thousands of parallel fine lines scribed into its reflecting surface. The details of how diffraction gratings work are outside the scope of this book (see Appendix III for sources of information on this topic), but anyone can see that they *do* work

FIGURE 4.5 A modern general-purpose astronomical spectrograph (FLAMES – Fibre Large Area Multi-Element Spectrograph) for ESO's VLT. The spectrograph is the collection of room-sized boxes in the foreground. The 8.2-m Kueyen telescope is in the background and gives an idea of scale. (Reproduced by kind permission of ESO/H.H.Heyer).

by holding a CD or DVD up to the light. The CD or DVD is likewise a reflector with thousands of parallel fine lines. Although the lines are much shorter than those of a diffraction grating, nonetheless a dazzling display of colours will be seen whenever the disk is held at the correct angle to the light.

By around 1980, spectrographs were being built that were capable of measuring a star's velocity with an uncertainty of just ±20 m/s. Potentially, therefore, such instruments could detect the motions of stars arising from exoplanets the mass of Jupiter and upwards and naturally many groups of astronomers started to try to do just that. The majority of these exoplanet-hunting spectrographs used the same operating principle as that of Huggins – the introduction into the star's spectrum of additional lines of known wavelength by passing the light through a source of those lines in front of the spectrograph. Huggins used an emission lamp, resulting in emission (bright) lines in the spectrum. The modern variants of his design superimposed absorption lines by passing the star's light through an enclosure containing hydrogen fluoride or iodine.

When and by whom the first exoplanet was detected is a matter of some debate. There are arguments for and against at least seven candidates – the planets around:-

γ Cep A (1988 – the Greek alphabet is listed in Appendix I for reference),
ε Eri (1992),
β Gem (1993),
PSR 1257+12 (1992 – two exoplanets),
PSR B1620-26 (1993) and
51 Peg (1995).

The first three of these were only tentative, unconfirmed (until much later) suggestions which in some cases were subsequently withdrawn. The fourth candidate has stood the test of time and was the first undoubted detection of exoplanets of any type. The fifth could equally well have been interpreted as a small star rather than an exoplanet and its true status (exoplanet) was not confirmed until 2003. The planets around pulsars, though, are likely to be very different in nature and origin from the exoplanets formed as part and parcel of their host star's formation (Chap. 13). What is certainly true, however, is that the planet detected around the star 51 Peg was the first confirmed discovery of what we now know to be a normal exoplanet orbiting a normal star.

The Real Thing – 51 Peg b

The epoch-making discovery of 51 Peg's planet was announced by Michel Mayor and Didier Queloz (Box 4.2) on the 5th October 1995 at a conference in Florence.

Box 4.2 Professors Michel Mayor and Didier Queloz

Michel Mayor
Born in Lausanne in 1942 Michel is currently a Professeur Honoraire and a past Director of the University of Geneva Observatory. He has degrees from the universities of Lausanne and Geneva and has been awarded the Marcel Benoist, Balzan, Shaw and Viktor Ambartsumian prizes and the Albert Einstein medal. He has been involved in the discovery of numerous exoplanets including the least massive one yet found at ≥ 1.9 Earth masses – Gliese 581e. His interests in very high resolution spectroscopy led him to the construction of the Elodie instrument with which 51 Peg b was discovered and the subsequent design and construction of the highly successful exoplanet hunting Coralie and High Accuracy Radial velocity Planet Searcher (HARPS) instruments.

Didier Queloz
Didier was 29 and studying for his Ph.D. with Michel Mayor as his supervisor when the discovery of 51 Peg b was made. His task was analysing the data from Elodie and he chose 51 Peg to study because it was one of the brightest in a list of 120 possible candidate stars. The Elodie spectrograph was newly commissioned on the 1.93-m telescope at the Observatoire de Haute Provence, when in November 1994 Didier started his campaign of observations. By March 1995 it was clear that the radial velocity of 51 Peg was changing regularly with a stable 4.2-day period. But further observations were needed to eliminate alternative explanations such as pulsations of the star or a spot on the star's surface. 51 Peg became observable again in July 1995 and within a couple of months Didier and Michel were convinced that it was a real exoplanet that was causing the changes in the star's radial velocity. After the announcement of the discovery in Florence in October 1995, Didier said in an interview for this book, that because of the unexpected closeness of the exoplanet to its host star (Figure 4.10):

> ... most of the people in that room didn't believe any word of what we said! This was so new that [it was] not only a planet but

a planet in a completely unexpected [*position*] and not predicted by the theory. [*Following*] 30 years of space missions, probes in the solar system, meteoritic studies and whatever you can … [*and now*] two guys from a small country with a new kind of instrument demonstrate one day in Florence that [*real exoplanets are*] not exactly what we have thought from the billions of dollars spent to study the planetary systems.

Didier is now Professor of Astrophysics at the University of Geneva and has been involved in the discovery of over a hundred exoplanets. He remains ecstatic, though, about that first discovery:

> I am one of the few people – the very, very few people that opened the door [*to a major scientific discovery*] – it changes a little bit the way you feel the sciences – you touch the very top – and not all scientists are lucky enough to reach that level – to touch such high level discoveries.

Away from the telescopes, Didier has many interests – "Like a real Swiss I ski a lot – I love to ski." With his three children he also likes being outside "I like nature, music – all kinds of music- … and I like also vacations (when I can get some!)" (Figure 4.6).

FIGURE 4.6 Michel Mayor (left) and Didier Queloz in 1998 during the installation of the Coralie spectrograph on the 1.2-m Euler telescope at ESO's La Silla site. (Reproduced by kind permission of D.Queloz).

From September 1994 to September 1995 they had measured the star's radial velocity using a spectrograph on the 1.93-m telescope at the Observatoire de Haute Provence. The spectrograph, named Elodie, was designed for very precise radial velocity measurements but was not of the hydrogen fluoride or iodine absorption cell type. Instead Elodie produced very stable high resolution spectra by being kept in a temperature-controlled room away from the telescope and by having the light from the star fed into it via fibre optic cables. A sophisticated analysis of the data then gave velocities to an accuracy of ±10 m/s.

By March 1995 Mayor and Queloz knew that they probably had caught the first normal exoplanet because the star's velocity was changing up and down by nearly 120 m/s every 4.2 days (Figure 4.7). But more observations were needed to eliminate other explanations for the changes such as stellar pulsations or star spots. After a nerve-wracking wait of 4 months 51 Peg became observable again in July and within a few weeks there was no doubt – it was a planet!

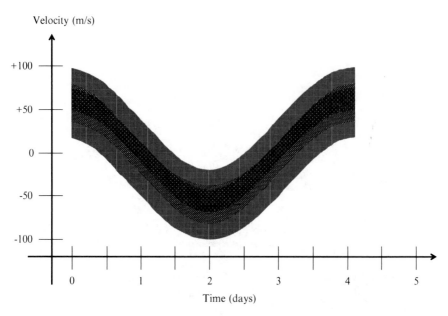

FIGURE 4.7 51 Peg velocity curve. The data taken over a period of a year have been plotted into a single cycle of 4.2 days so that the variation can clearly be seen. The width of the curve reflects the scatter (uncertainties) in the data (After M. Mayor, D. Queloz, *Nature* **378**, 355, 1995).

Many of the astronomers at the Florence conference though were not so convinced because the planet was so unexpectedly close to its host star. But just a few days after the conference, the planet's existence was confirmed by Geoff Marcy and Paul Butler (Box 4.3) using the Hamilton spectrograph on the Shane 3-m telescope of the Lick observatory. This spectrograph did use an iodine absorption cell to achieve its ±3 m/s accuracy.

Box 4.3 Prof. Geoff Marcy and Dr. Paul Butler

Geoff Marcy

Geoff and his team discovered 70 of the first 100 exoplanets to be found and he is still the most prolific planet hunter around. He was born in 1954 in Michigan and studied magnetic fields in solar-type stars to obtain his doctorate in 1982 from the University of California, Santa Cruz. He is currently Professor of Astronomy at the University of California at Berkeley. Amongst many honours he shared the Shaw prize with Michel Mayor in 2005 and was awarded the Carl Sagan prize in 2009. His ambitions for the future are to continue to discover and characterise planets beyond the solar system.

Regarding his crucial confirming observations of Mayor and Queloz' discovery of 51 Peg b in 1995, he says "We confirmed that planet around 51 Pegasi within 10 days at the 120-in. telescope, setting off a firestorm of media coverage, including interviews by every major television and newspaper."

He now thinks "... the greatest scientific question is the typical distance separating technological civilizations in the Milky Way Galaxy. It could be 10 or 10,000 light years, or more." While for science in general he is "... interested in ensuring equal treatment of everyone in science, independent of gender, race, and sexual orientation."

"Our greatest human challenge is to promote the survival of our species." he says.

He is married to Dr Susan Kegley who is Principal and CEO of the Pesticide Research Institute in Berkeley, California and

(continued)

Box 4.3 (continued)

FIGURE 4.8   Prof. Geoff Marcy (Reproduced by kind permission of C. Rose).

he loves playing tennis; "(I) especially enjoy court time with the UC Berkeley men's and women's tennis teams" (Figure 4.8)

Paul Butler
Paul is currently a staff member of the Carnegie Institution's Department of Terrestrial Magnetism. His research interests continue to involve exoplanets but also include Sun-like stars, supergiants and Cepheid variable stars. Between 1985 and 1989 he collected a B.A. in Physics, a B.Sc. in Chemistry and an M.Sc. in Physics from San Francisco State University. He topped this lot off in 1993 with a Ph.D. in Astronomy from the University of Maryland. He has also picked up the Bernard Oliver Memorial award from the Extrasolar Planetary Foundation, the National Academy of Science's Henry Draper medal and the Carl Sagan Memorial award as well as being *Discover* magazine's space scientist of the year in 2003.

Once the fuss arising from the discovery was over, further observations soon established the details of the exoplanet and its star. 51 Peg is a star similar to but slightly larger, slightly more massive and slightly cooler than the Sun and it is 50 light years away from us. From a good observing site it can just be seen with the naked eye about halfway between α and β Peg and slightly to the West of the line joining those two stars (Figure 4.9). The exoplanet discovered in orbit around it has about half the mass of Jupiter, but is so close to its star that its cloud top temperature reaches around 1,100–1,200°C (hot enough to melt copper). The high temperature means that the exoplanet is probably a bit larger than Jupiter despite its smaller mass. The separation of 51 Peg a (the star) and 51 Peg b (the planet) is just 0.05 AU (Figure 4.10) and it completes its orbit in the 4.2 day period of the star's velocity variations (Jupiter for comparison at 5.2 AU from the Sun takes 11.9 years to complete an orbit). The planet is so close to the star that the star's gravitational field causes it always to keep the same face towards the star. The planet therefore also rotates once every 4.2 days.

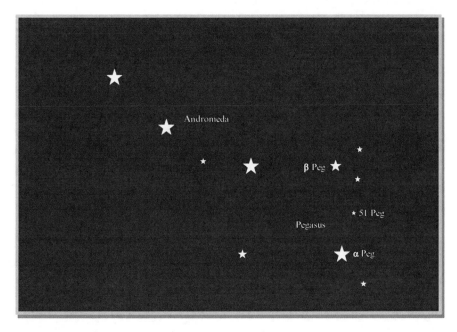

FIGURE 4.9 The position of 51 Peg in the sky.

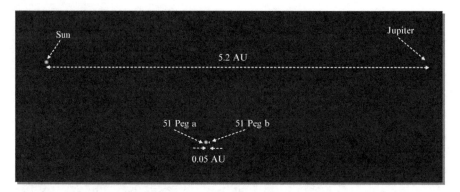

FIGURE 4.10 Comparison of the Sun and Jupiter with the 51 Peg system. Relative to the separations of the stars and planets, the sizes of the stars are exaggerated by a factor of 10, those of the planets by a factor of 60.

51 Peg b is not only the first exoplanet of a normal star but is the first of the hot Jupiters. Exoplanets that are very close to their host stars are the rule rather than the exception – half of those so far discovered are closer to their stars than the Earth is to the Sun. The nature of 51 Peg b should not though have caused the surprise that it did (some astronomers at the time dubbed it a 'crazy' planet) since the radial velocity method of detecting exoplanets is biased towards finding massive planets close to low mass stars. The reason for this bias is just that massive planets (and low mass stars) in close mutual orbits produce the largest velocity variations in the shortest times for the star and so are the easiest to detect.

The mass of 51 Peg b is now known to be at least 0.46 Jupiter masses, however it could be quite a bit more than that. The reason for this uncertainty is that we do not know the angle of the exoplanet's orbit to the line of sight. For exoplanets detected via the transit method we know that the planet's orbital plane has to be very close to being perpendicular to the plane of the sky (90° inclination) so that we are looking almost along the plane of the planet's orbit. However for other methods of detecting exoplanets, including the radial velocity method, only rarely can the orbital inclination be estimated – in fact it is only known for about 20% of exoplanets.

The radial velocity method (see Chap. 5) as its name implies, detects exoplanets via the host star's radial velocity changes. If the plane of the exoplanet's orbit is at an angle to the line of sight, then so will be the plane of the star's orbit. The radial velocity of the star that is measured will then only be a component of the true orbital velocity of the star. If the star's orbital plane is inclined at 45° to the line of sight, for example, then the true orbital velocity will be 40% higher than the measured velocity and the true mass of the exoplanet will be 40% higher than the minimum (Appendix IV). Statistically, though, the true mass of an exoplanet whose inclination is unknown is likely to be around 25% higher than the minimum value. The minimum exoplanet masses obtained via the radial velocity method are thus probably quite close to the true masses for many exoplanets.

Near Misses – γ Cep A, ε Eri and β Gem

In 1979 Bruce Campbell and Gordon Walker of the University of British Columbia devised a spectrograph able to measure the radial velocity of a star to an accuracy of ±15 m/s. Until then the best that most spectrographs could achieve was around ±1,000 m/s and some instruments gave far poorer results than that. The improvement in accuracy was achieved, as mentioned earlier, by passing the star's light through a container of a gas before it entered the spectrograph. The gas absorbed light at its own characteristic wavelengths, adding a series of absorption lines to the star's spectrum. These artificially produced lines have known wavelengths and so the wavelengths of the star's lines can be found by comparing the positions of the two sets of lines. Campbell and Walker used hydrogen fluoride, which when heated to about 100°C, produces a series of sharp absorption lines in the near infrared (Figure 4.11).

For the next 6 years Campbell and Walker along with Stephenson Yang used their spectrograph on the 3.6-m Canada-France-Hawaii telescope (CFHT) to monitor the radial velocities of 29 stars, and published preliminary results for 16 of their programme stars in 1988. Seven of these stars showed significant velocity variations. The radial velocity of γ Cephei in particular decreased from around +750 to –750 m/s over the period. This large change in velocity was attributed to the presence of a second star – now known probably to be a red dwarf with a mass around 0.4 solar masses. The original star is thus called γ Cep A and its

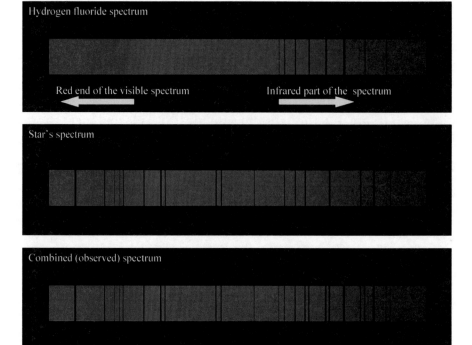

Figure 4.11 The principle of Campbell and Walker's radial velocity spectrograph. (NB – for clarity, the separation of the hydrogen fluoride lines has been much exaggerated. Also the star's lines are schematic only).

stellar companion γ Cep B. However the large steady trend in the radial velocity due to the companion star had a small ripple superimposed upon it. This ripple had an amplitude of 25 m/s and a period of about 2.7 years. The team attributed the ripple to the possible presence of an exoplanet with a minimum mass around 1.7 Jupiter masses and went on to add:

> Since we have identified a period to the radial velocity variations … . This star therefore has the firmest evidence for a very low mass companion.
>
> (Astrophysical Journal **331**, 902, 1988)

However in the conclusion to their paper, they wrote:

> For seven of the 15 stars there is a 'possible' or 'probable' companion in the range ~1 – 9 Jupiter masses, assuming periods less than 50 years. … This suggests that the seven companions we have tentatively detected might be more closely related to planets than

to brown dwarfs ... Additional information is required before a firm conclusion on the nature of these companions can be reached.

By 1992 further observations had been made, but these did not support the earlier ideas and the claim that γ Cep A had a planetary companion was withdrawn. The velocity changes instead were attributed to the rotation of γ Cep A itself:

> While binary motion induced by a Jupiter-mass companion could still explain the periodic residuals, γ Cep is almost certainly a velocity variable yellow giant. ... The λ8662 Ca II emission line [*Author's note – this is a bright spectrum line in the infrared at a wavelength of 866.2 nm or 8662 Ångstroms which is produced by calcium atoms that have lost one of their normal entourage of electrons*]... varies in phase with the 2.52 year period which ... strongly implies that it is in fact the star's period of rotation.
> (G.A.H Walker et al., Astrophysical Journal, **396**, L91, 1992)

In 1993, the team with other collaborators found a periodicity of 1.6 years in the radial velocity of β Gem (Pollux) whose amplitude was 46 m/s. They concluded though that:

> With our present data, the rotation [*of the star*] – versus – revolution [*of an exoplanet*] question cannot be resolved for β Gem.

Similarly an inconclusive radial velocity variation of 15 m/s over about a 10-year period was found for ε Eri in 1995.

Thus none of these three stars had definitely been identified as possessing an exoplanet when Mayor and Queloz announced their discovery of 51 Peg b. Subsequently however, exoplanets have been found for all the stars. In 2000, a 0.86 Jupiter mass planet in a 6.9-year orbit was found by Artie Hatzes et al. orbiting ε Eri. Three years later Hatzes and his collaborators found γ Cep A's exoplanet. It is in a 2.48 year orbit, has a minimum mass of 1.7 Jupiter masses and is 2.13 AU out from its host star. Finally in 2006, Hatzes' team found a 2.3 Jupiter mass exoplanet in a 1.6-year orbit, 1.6 AU out from β Gem.

In retrospect, it is clear that, at least for γ Cep and β Gem, their exoplanets *had* marginally been detected by the earlier work. Should the discovery of the first exoplanet therefore be given as γ Cep Ab in 1988? The answer, most scientists would agree, must be 'No'. The question of precedence in scientific discoveries has a long and acrimonious history. To quote just one of thousands of examples; before its discovery, Neptune's position in the sky had been predicted independently by John Adams and Urbain Le Verrier. Adams' prediction pre-dated that of Le Verrier, but it was Le Verrier's prediction that was used by Johann Galle in 1846 to discover the planet. So who gets the credit for the prediction of Neptune's position? Most scientists (the question is complicated by chauvinism since Adams was British and Le Verrier, French) give the credit to Le Verrier since he not only published the prediction but had it independently confirmed (by, in this case, the actual discovery of Neptune). Thus in modern science the criteria for being credited with a discovery are:

1. A clear statement of the nature of the discovery that is available in the public domain (nowadays this is often a press release or announcement at a conference, followed by a publication in a peer-reviewed journal)
2. Adequate supporting evidence for the discovery and
3. Independent confirmation

Thus although the γ Cep and β Gem exoplanets were eventually confirmed, there was no clear statement of their discovery following the original work (and for γ Cep A, what statement there had been was specifically withdrawn). Furthermore the evidence in both cases was equivocal – even the discoverers were unsure whether the velocity changes arose from exoplanets or stellar rotations or other causes.

The Very First Exoplanets – PSR 1257±12 B and PSR 1257±12 C

51 Peg b caused a great deal of surprise because of its proximity to its host star. The first definitive exoplanets to be found, though, were even more astonishing, for they were orbiting a pulsar. Pulsars (PULSAting Radio Sources) are the remnants of those extreme stellar explosions called supernovae. Supernova explosions destroy stars that are several times more massive than the Sun – so how could mere planets hope to survive such holocausts? Before answering that question though, let us see what evidence there is that there *are* exoplanets orbiting pulsars.

Pulsar exoplanets are detected through the Doppler shift but operating in a slightly different fashion from the way that it allowed

51 Peg's exoplanet to be found. A pulsar emits its radio pulses at very regular intervals. The pulsation period is typically constant to between one part in 100,000,000,000 (=10^{11}) and one part in 10,000,000,000,000,000 (=10^{16}) – a clock that kept time as well as such pulsars would gain or lose just one second in 3,000 years in the first case and need *300 million* years to do so in the second case! – and that's as good as the best modern atomic clocks can do.

When a pulsar has a planet orbiting around it, then as in the case of a normal star, the actual situation is that the pulsar and exoplanet both orbit around their common centre of gravity (Figure 4.3). Unless the orbits are perpendicular to the line of sight, then in one part of its orbit the pulsar will be approaching the Earth, and in the opposite part it will be going away from us.

To see how this enables the exoplanet to be detected, imagine that the pulsar's orbital speed is 1 m/s and that its orbital plane is along the line of sight. Imagine also that its pulses are emitted at exactly one second intervals. Take two successive pulses emitted during that part of the orbit when the pulsar is approaching us directly (Figure 4.12). The pulsar will be 1 m closer to us when the

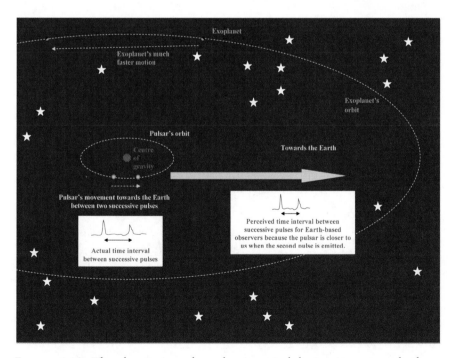

FIGURE 4.12 The shortening of a pulsar's period during its approach phase towards the Earth (not to scale).

second pulse is emitted than it was when the first pulse came out. The radio waves of the second pulse will therefore have 1 m less distance to travel to reach us. Since radio waves (like light) move at 300,000,000 m/s, this means that we will receive the second pulse 1/300,000,000 s (= 0.000.000,003 s or 3×10^{-9} s) earlier than we would have done if the pulsar had been stationary. Since we have assumed that the two pulses were separated by exactly one second when they were emitted, we will thus actually receive them separated by an interval of 0.999,999,997 s. Likewise when the pulsar is moving directly away from us, the perceived period will be 1.000,000,003 s. This may seem like a very small change and in fact it means that the stability of the pulses is still three parts in 1,000,000,000 (= 3 parts in 10^9 – 1 s in 10 years), but the stability of the emitted pulse interval is at least 300 times better than that. Thus the change in the pulse period is relatively easily observable and from it the presence of the exoplanet can be inferred.

In 1991 Aleksander Wolszczan (Box 4.4) was using the 305-m Arecibo radio telescope to observe a pulsar in Virgo. The pulsar, labeled PSR B1257+12 from its position in the sky in 1950 (Figure 4.14), lies about 1,000 light years away from us and its pulse period is 0.0062 s (6.2 ms). Dale Frail (Box 4.4) also observed the pulsar using the Very Large Array (VLA) radio telescope in

---

Box 4.4 Professors Aleksander (Alex) Wolszczan and Dale Frail

Alex Wolszczan

Alex is currently the Evan Hugh Professor of Astronomy and Astrophysics at Penn State University, being appointed to the chair in 1998. He was born in Szczecinek, Poland, some 200 miles North-West of Warsaw in 1946 and obtained his degrees (M.Sc – 1969, Ph.D. – 1975) from the Nicolaus Copernicus University in Toruń. Since 1982, he has been based in the U.S.A., though until recently also holding a part-time professorship at the Nicolaus Copernicus University and he continues to be a member of the Polish Academy of Sciences. Amongst other honours, he has won the American Astronomical Society's Beatrice

M. Tinsley prize, the Polish Physical Society's Marian Smolu-chowski Medal and in 2002 featured on a Polish stamp celebrating the nation's achievements in the second millennium.

Dale Frail

Dale, a Canadian by birth, obtained his Bachelor's degree from Acadia University, Nova Scotia in 1983 and his Doctorate from the University of Toronto, Ontario, in 1989. He currently holds the post of Assistant Director, Science and Academic Affairs at the NRAO. His research interests include gamma-ray bursts as well as pulsars and their exoplanets. Amongst other honours he has recently been awarded Guggenheim Fellowship.

> "… we were not looking for planets at all." Dale explains. "It is really a tale of two telescopes: Alex at the 305-m in Arecibo and me at the Very Large Array (VLA) in Socorro, New Mexico. Alex had made the discovery of PSR B1257+12 with the Arecibo telescope and he had started accurately measuring the arrival times between successive radio pulses from the pulsar. He saw something he did not understand and he called and asked me to measure a precise position for the pulsar. The VLA, as an interferometer, can get a much better position than a single dish telescope like Arecibo. After a few attempts I succeeded in getting an accurate position for the pulsar and passed that position to Alex. The discovery of the first two planets followed almost immediately."

Despite being the first to be found, pulsar planets are rare. Dale thinks that their real significance lies in the encouragement that the discovery gave to other exoplanet hunters "Back in the early 1990s the searches around normal stars had been going on for some time by a small number of groups. The technique was difficult but slowly improving. Funding was difficult and telescope time was hard to get. Some people thought it all a waste of time. I remember clearly the excitement with which these early planet searchers received the publication of Wolszczan & Frail. The attitude was 'what we are attempting is not in vain. If planets can be found in orbit around something as bizarre as a pulsar, surely we will be successful.'" (Figure 4.13)

(continued)

Box 4.4 (continued)

FIGURE 4.13 Prof. Dale Frail (Reproduced by kind permission of D. Frail).

New Mexico in order to determine its position very precisely. The two observers found that over several months the pulse period varied by up to 0.000,000,000,03 s in complex way. Although the period variations were not the simple curve (Figure 4.14) that would result from a single exoplanet, Wolszczan and Frail quickly realized that they resulted from *two* exoplanets in slightly different period orbits around the pulsar. The exoplanets, called PSR B1257+12 B and PSR B1257+12 C, are in nearly circular orbits with radii of 0.36 AU and 0.46 AU, respectively (similar sized orbits to that of Mercury). Their orbital periods are 66.5 and 98.2 days, their minimum masses 3.4 and 2.8 Earth masses and they are likely to have surface temperatures around 400°C – comparable with that of the day side of Mercury.

Don Backer et al. confirmed the discovery in July 1992 with observations made from the National Radio Astronomy Observatory (NRAO) at Green Bank and so PSR B1257+12 B and PSR B1257+12 C became on the 9th January 1992, when the discovery

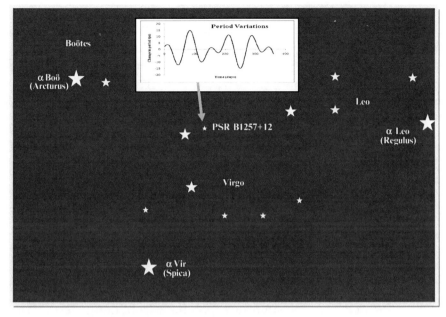

FIGURE 4.14 Position of PSR B1257+12 in the sky and a schematic plot of the variations in the pulsar period over a few months.

paper was published in *Nature*, the first ever planets to be found without any doubts beyond the solar system. The presence of a third, lower mass, planet was also suspected from the data on the pulsar's period. This exoplanet, labeled PSR B1257+12 A, was confirmed by Wolszczan in 1994. It is by far the least massive exoplanet found to date with a minimum mass about twice the mass of our Moon (0.015 Earth masses) and it circles the pulsar in a 0.19 AU orbit every 25.3 days.

The possible existence of further planets within the system have been suggested, such as a 100 Earth mass gas giant at 40 AU or a 0.0004 to 0.004 Earth mass dwarf planet 2 to 3 AU out, but neither of these proposals has been confirmed and indeed the very small pulse period variations that suggested their existences probably arise from other causes.

In 2003, Wolszczan working with Maciej Konacki determined the inclinations of the exoplanets' orbits. PSR B1257+12 B's orbit is inclined at 53° ±4° to the plane of the sky, while PSR B1257+12°C's orbit is inclined at 47° ± 3°. Knowing the orbital inclinations enables the exoplanets true masses to be determined, not just lower limits to those masses. Assuming the pulsar itself has a mass of 1.4 solar masses (the Chandrasekhar limit – the mass at which a white dwarf will collapse to form a neutron star), then the two planets have actual masses of 4.3 ±0.2 Earth masses (PSR B1257+12 B) and 3.9 ±0.2 Earth masses (PSR B1257+12 C).

Since it is now clear that there are at least three smallish exoplanets are orbiting PSR B1257+12, we come back to the question of how they could have survived the supernova explosion in which the pulsar originated. Two possibilities are proposed. The first is that the exoplanets at one time were gas giants with Jupiter's mass or more. During the star's explosion, the outer layers of those giant planets were stripped away, but their rocky central cores survived. Planets formed in this way are sometimes called Chthonian (pronounced 'thonian') planets from the Greek, *chthon*, meaning pertaining to the Earth. The second possibility is that the planets formed after the supernova, perhaps from debris left by the explosion. Though this may seem unlikely, Type Ia supernovae originate within a close binary system containing a white dwarf and a main sequence or sub-giant star. The explosion occurs on the white dwarf and the companion star is likely to be blasted free during the explosion, but some of its material might remain trapped around the neutron-star remnants of the white dwarf and so then be available to form planets.

Methuselah – The Oldest of Them All – PSR B1620-26 b

From 1992 onwards Don Backer, Stephen Thorsett, Steinn Sigurdsson, their co-workers and others were suggesting that the timings of the pulse period of a pulsar/white dwarf binary in the globular cluster, M4, implied the presence of a 'low mass companion' within the system. Thorsett et al. went as far in 1993 as to postulate that it might be a planet in a 10 AU orbit or a small star in a 50 AU orbit. The pulsar, PSR B1620-26 A has a period of 11 ms and is about 7,000 light years away from us. Its mass, thought to be 1.35 times that of the Sun, like that of most pulsars is close to the Chandrasekhar limit. The white dwarf, WD B1620-26 or PSR

B1620-26 B, has a mass of 0.35 solar masses and is in a 0.8 AU orbit around the pulsar with a period of just over 6 months. Sigurdsson et al. finally obtained direct images of the white dwarf using the HST in 2003.

Like the cases of γ Cep A, ε Eri and β Gem, the early indications of the presence of an exoplanet took a long time to confirm. Even in 1999, Thorsett and his team could only claim

> We find that the second companion most likely has a mass m ~ 0.01 solar masses - it is almost certainly below the hydrogen-burning limit (m < 0.036 solar masses, 95% confidence) – and has a current distance from the binary of ~35 AU and orbital period of order 100 year.
>
> (Astrophysical Journal **523**, 763, 1999)

Uncertainties still remain about the exoplanet, but it may now be regarded as having had its existence confirmed. It is currently thought to orbit the pulsar/white dwarf binary at a distance of about 23 AU with an orbital period of around a century. Its mass probably lies between 1.5 and 3.5 times that of Jupiter.

The globular cluster, M4, is thought to have originated around 12.7 billion years ago (just 1 billion years after the Big Bang) so this is also the moment of birth for the original objects that are now the pulsar, white dwarf and exoplanet. The prodigious age of the exoplanet – three times older than the Earth – making it easily the oldest exoplanet found so far has led to it receiving the unofficial name of 'Methuselah' (after the oldest person mentioned in the Old Testament – he reputedly lived until he was 969). Officially though, the exoplanet is called PSR B1620-26 b (sometimes PSR B1620-26 c).

Unlike the planets of PSR B1257+12, Methuselah probably formed in a normal fashion and it and its host star (now the white dwarf) were gravitationally captured by the neutron star after it had been formed in the supernova explosion. Stars in globular clusters are tightly packed and close passages and interactions between them are likely to be common, especially in the dense central regions. Originally therefore Methuselah would have been a normal planet of a smallish but normal main sequence star. What is now the neutron star would then have been a more massive main sequence star forming a close binary system with a second, smaller normal star. In due course the neutron star precursor would finish its main sequence life, evolve into a giant and then into a white dwarf, still retaining its normal companion star. Eventually the companion star would also start to evolve towards becoming a giant (less massive stars have longer main sequence lifetimes than more massive stars). As the companion's size increased, material would flow from it to be captured by and accumulate on the surface of the white dwarf. When the white dwarf's mass exceeded the Chandrasekhar limit of 1.4 solar masses it collapsed to become a neutron star, so initiating the supernova explosion. The explosion did not disrupt the binary system, although it is likely that the companion star would have lost some of its mass. After the explosion the companion star continued its evolution and in turn ended as a white dwarf, leaving a neutron star-white dwarf binary system.

Sometime later a close encounter of this binary system with Methuselah and its host star led to the neutron star capturing the latter pair whilst simultaneously the white dwarf was ejected to become a solitary star (though probably still retained within the globular cluster). Methuselah's host star, now a close companion to the neutron star, eventually expanded in size as it also evolved away from the main sequence. This time the mass exchanged onto the neutron star would cause it to spin more and more rapidly, until it achieved its present rotation rate of five and a half thousand rpm. Finally Methuselah's original host star would complete its evolution to a white dwarf, leaving the system as we now see it and with Methuselah orbiting the binary star pair at a considerable distance from them.

That is probably not the end of the story, however. At some time in the future, the complex path of PSR B1620-26 within M4 is likely to take it back towards the centre of the globular cluster. In those dense central parts further gravitational encounters with stars are to be expected. Since Methuselah is now relatively weakly gravitationally bound to the neutron star and the white dwarf, it could easily be ejected during such an encounter, thus finally becoming a free-floating solitary planet within M4, or even being ejected entirely from the globular cluster.

5. On the Track of Alien Planets – The Radial Velocity or Doppler Method (~70% of All Exoplanet Primary Discoveries)

The successful detection of exoplanets through the host stars' velocity changes had to await the development of spectrographs capable of measuring velocity changes of a few metres per second. Although 51 Peg b and some other exoplanets have been discovered using relatively conventional spectrographs, most exoplanets have been found using spectrographs equipped with absorption cells.

In a conventional spectrograph, the wavelengths of the star's lines are found by comparing their positions along the spectrum with those of emission lines produced by an artificial and local source and whose wavelengths are known accurately from laboratory studies (comparison spectrum – Figure 5.1). However, although the emission lines' wavelengths may be accurately known, the light from their source (usually a gaseous emission lamp) does not follow exactly the same optical path through the instrumentation as the light from the star, with a consequent lose of precision in determining the wavelengths of the star's lines. However, by making the light from the star pass through a transparent container of an absorbing gas (an absorption cell) that is positioned between the telescope and the spectrograph, artificial absorption lines will be added to the star's spectrum that *have* followed the same optical path through the spectroscope as the star's light.

Early designs of absorption-cell spectrographs used hydrogen fluoride (HF) gas as the absorbing medium. HF though is lethally corrosive, reacts chemically with glass, has to be heated to 100°C and needs a cell at least a metre in length.

C. Kitchin, *Exoplanets: Finding, Exploring, and Understanding Alien Worlds*, Astronomers' Universe, DOI 10.1007/978-1-4614-0644-0_5,
© Springer Science+Business Media, LLC 2012

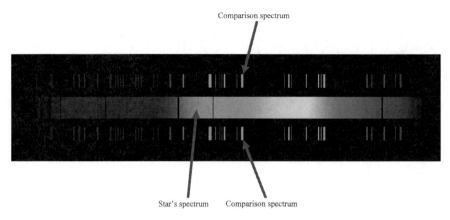

FIGURE 5.1 Schematic and simplified appearance of the stellar and comparison spectra for a conventional spectrograph.

In designing an absorption cell for use on the Hamilton spectrograph of the Lick Observatory's 3-m Shane telescope in the early 1990s, Marcy and Butler therefore chose to use molecular iodine in place of the HF. Although nasty in its gaseous form, iodine is non-lethal and less corrosive than HF. Furthermore its much stronger absorbing power meant that a cell only 100 mm in length would be needed and the iodine gas pressure could be just 1% of that of the atmosphere. Finally the iodine cell operates at lower temperatures (around 50°C). The disadvantage of molecular iodine is that it has many thousands of absorption lines. These blend with the lines from the star so that special techniques are needed to analyze the data. Despite the latter problem, this was the instrumental arrangement that Marcy and Butler used to confirm the discovery of 51 Peg's exoplanet in 1995 and later to discover many more exoplanets of their own including 70 of the first 100 to be found.

The use of iodine absorption cells has now spread widely amongst exoplanet hunters. Marcy and Butler continue to be at the forefront of the discoveries, but many large telescopes now have high resolution spectrographs which include the option of using an iodine cell. These include the 10-m Keck telescopes, the 11-m Hobby Eberly telescope, the 8.2-m Subaru telescope, the 6.5-m Magellan telescopes and – one of the most successful with over 30 exoplanet discoveries to its credit – the 3.9-m Anglo-Australian telescope.

The designs of conventional spectrographs have been improved so that their accuracies now rival those of the iodine-cell instruments. Indeed, as we have seen, the discovery of 51 Peg's exoplanet was made using a relatively conventional spectrograph (Elodie). The wavelengths of the star's spectrum lines are found from comparison lines that are usually placed to one side or sometimes both sides of the star's spectrum (Figure 5.1). Almost all exoplanet-hunting spectrographs use a thorium-argon hollow cathode lamp to produce the comparison lines.

In order to reach accuracies in Doppler shift measurements of a few metres-per-second with a spectrograph of conventional design painstaking and meticulous care needs to be taken at all stages of the process. Foremost amongst the requirements is to remove the spectrograph from the vicinity of the telescope and house it separately in a temperature-stabilized room. The light from the star is then fed from the telescope to the spectroscope by a fibre-optic link. The huge increase in stability that results from this move arises because spectrographs contain many optical components that need to be maintained in very precise positions with respect to each other if the spectroscope is to work optimally. By operating at a constant temperature any effects upon the relative positions of the optical components due to thermal expansion or contraction are eliminated. Also the optical properties of some optical components, such as prisms, vary with temperature so this problem is also eliminated. Finally the gravitational loading and the resultant stresses and distortions within the apparatus is constant and can be corrected – unlike the case when the spectrograph is mounted on the telescope and moves around to different orientations as the telescope tracks the object being observed.

Foremost amongst these conventional exoplanet-hunting spectrographs was Elodie. This was designed and built by André Barranne, Mayor, Queloz and their associates and employed by Mayor and Queloz to discover 51 Peg's exoplanet. It was used, amongst other purposes, on the 1.93-m telescope of the Observatoire de Haute Provence from 1993 to 2000 for the Northern Extrasolar Planet Search programme, discovering over 20 exoplanets in addition to 51 Peg b. A very similar instrument, Coralie, operates in the southern hemisphere on the 1.2-m Leonard Euler telescope at ESO's La Silla observatory conducting the Southern

Sky Extrasolar Planet search programme from 1998 onwards. Coralie eventually achieved an accuracy in measuring radial velocities of ±2 m/s. Between them, Elodie and Coralie either acting individually or through combining their observations have made 60 exoplanet discoveries to date. Elodie was replaced by SOPHIE (Spectrographe pour l'Observation des Phénomenes sismologique et Exoplanétaires) in 2006 – a spectrograph of very similar design to Elodie but with improvements such as a larger CCD detector with smaller pixels, higher optical efficiency, higher spectral resolution, etc. and it can now also reach precisions of ±2 m/s. Observations using SOPHIE, sometimes combined with the Elodie archive data, have so far revealed another 11 new exoplanets.

The ESO's La Silla Observatory is also home to the HARPS (High Accuracy Radial velocity Planetary Search) exoplanet-hunting spectrograph in addition to Coralie. HARPS, commissioned in 2003, like Coralie, is housed in a temperature-controlled room, contained within a vacuum chamber and uses a thorium-argon comparison spectrum. It is fed by fibre-optic cables from ESO's 3.6-m telescope and achieves a radial velocity accuracy of around ±1 m/s. HARPS-NEF (HARPS – New Earth Facility) is a comparable instrument currently under construction to use on the 4.2-m William Herschel Telescope in the Canary Islands, so providing Northern-Hemisphere coverage. A laser comb comparison spectrum (Chap. 11) is currently being developed for HARPS – potentially improving the accuracy of its radial velocity measurements significantly.

A very recent development that holds out the promise of being able to study fainter stars than is currently possible and also many of them in a single observation whilst simultaneously being simpler and cheaper to construct than those instruments just discussed is the Exoplanet Tracker. In 2005, whilst operating on the 0.9-m Kitt Peak telescope, the instrument discovered its first exoplanet – a ≥0.49 Jupiter mass object in a 4 day orbit around a young star, HD102195 in Virgo. Exoplanet Tracker still uses a spectroscope, though one of relatively low dispersion compared with Elodie and HARPS and their like. The spectroscope is linked to an interferometer (interferometers are discussed in Chap. 11) and it becomes relatively straightforward to determine Doppler shifts to precisions of a few metres per second. For calibration purposes,

Exoplanet Tracker uses an iodine cell. It is likely that instruments of this design will be much more widely used in the future.

Can there possibly be a role for amateur astronomers in a field where the professionals casually play with multi-metre tele-scopes and massive spectrographs for nights on end to discover their exoplanets? A group of very professional amateurs in the Western U.S.A. is determined that the answer to that question will be 'Yes'. Spectrashift (http://www.spectrashift.com/index.shtml), led by Tom Kaye, has already managed to obtain spectra of τ Boö that clearly show the Doppler shifts induced by the presence of its ≥3.9 Jupiter mass exoplanet (discovered in 1996 by Marcy and Butler) using a 0.4-m off-the-shelf Meade™ telescope and an optical-fibre-fed home-built spectrograph. The group is currently building a 1.1-m telescope and associated spectrograph in South-Eastern Arizona with the firm intention of joining the exoplanet discoverers in due course. The group also intends to look for exoplanets via the transit approach.

6. On the Track of Alien Planets – The Transit Method (~23% of All Exoplanet Primary Discoveries)

Transits, eclipses and occultations are all essentially the same phenomenon. They are events when one astronomical object passes in front of another. During an eclipse the two objects are of comparable angular sizes – like the Moon eclipsing the Sun (Figure 6.1). In an occultation the distant object is angularly small compared with the nearer one – like the Moon occulting a star, whilst for a transit the situation is reversed – like one of Jupiter's satellites being silhouetted against the disk of the planet.

Transits have had a long history of being of interest to astronomers. To begin with, that interest lay in measuring the distance between the Earth and the Sun (the Astronomical Unit). When Mercury or Venus transits the Sun, observations of the transit from two well-separated spots on the Earth combined with simple trigonometry theoretically enables the Sun's distance to be found.

Transits of Venus gave the best hope of measuring the astronomical unit since Venus is much closer to the Earth than Mercury when it transits the Sun, but they only occur four times every 243 years – at intervals of 8, 121.5, 8 and 105.5 years. The last transit occurred on 8th June 2004 (Figures 3.2 and 6.2), and the next will be on 6th June 2012. In the eighteenth century epic voyages were made to set up observing sites as far as possible from European observatories so that as long a base line as was practicable was obtained. Captain Cook's first round-the-world voyage, for example, enabled observations of the 1769 transit to be made from Tahiti. Unfortunately, the values obtained for the astronomical unit from Cook's and other expeditions were of low accuracy

C. Kitchin, *Exoplanets: Finding, Exploring, and Understanding Alien Worlds*, Astronomers' Universe, DOI 10.1007/978-1-4614-0644-0_6,
© Springer Science+Business Media, LLC 2012

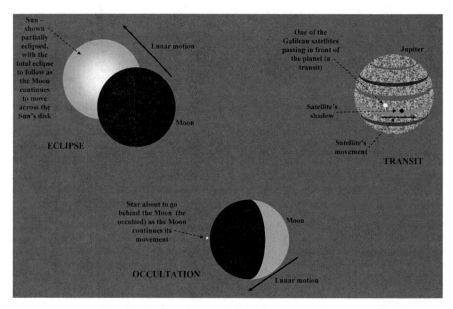

FIGURE 6.1   Eclipses, transits and occultations.

because the high contrast between the black silhouette of Venus and the bright photosphere of the Sun rendered their observations of poor quality.

Eighteenth and nineteenth century astronomers were interested in measuring the position of Venus with respect to the Sun in order to determine the distance between the Earth and the Sun. However as can be seen in Figure 6.2, during the transit, the planet obscures a small part of the radiation coming to us from the Sun – typically the brightness of the light we get from the Sun decreases by about 0.1% during a transit – so had they been interested, those earlier astronomers could also have tried to measure that change in luminosity. Success in that measurement would have produced a graph showing a slight dip in the Sun's brightness as Venus moved across its disk (Figure 6.3).

Modern astronomers now undertake exactly the same type of observations, not of Venus and the Sun, but of distant stars as their exoplanets transit in front of those stars' disks. The transit of Venus, however, as seen from the Earth, exaggerates the change in brightness because we are so close to the planet. A distant ET astronomer looking at a transit of Venus would see a brightness

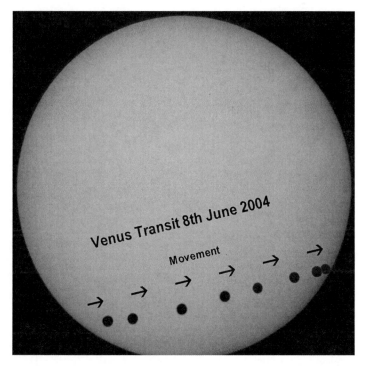

FIGURE 6.2 The transit of Venus across the Sun on the 8th June 2004. Images obtained at several stages throughout the transit have been combined to show the motion of the planet across the face of the Sun. (Copyright © C. R. Kitchin 2004).

change in the Sun of only 0.008%. That alien astronomer though would see transits every 225 days – the long and irregular intervals between Venusian transits as seen from the Earth arises from the two planets' orbital motions and the angle (3.4°) between their orbits.

With equipment similar to that which we have today, the alien astronomer would be pushed to detect a transit of Venus because the change in the Sun's brightness is so small. He / she / it though would be much more likely to pick up a transit by Jupiter or Saturn. The change in the Sun's brightness would then be by 1% and 0.8% respectively and even Uranian or Neptunian transits would cause solar brightness changes 16 times that of Venus.

Observing one transit is insufficient to count as a discovery of an exoplanet – many other processes can cause similar changes

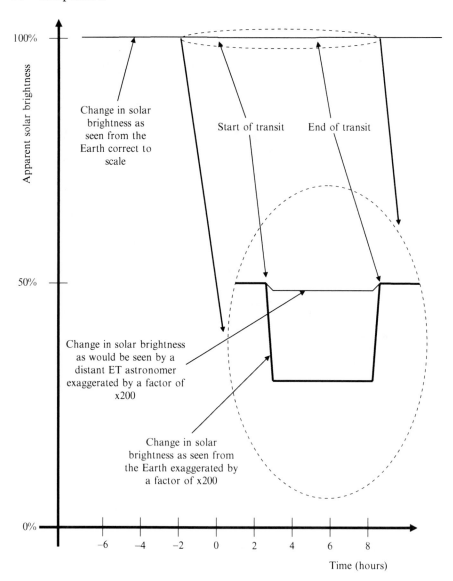

FIGURE 6.3 Schematic graph of the variations in the solar brightness during a transit of Venus.

to a star's brightness. A variation in a star's brightness that looks as though it might have been due to an exoplanetary transit, but which is due to some other process is called a false positive. Predominating amongst the causes of false positives are eclipsing binary stars, especially where the eclipse is a grazing one or where the eclipsing binary is so closely aligned with a foreground star that the two cannot be seen separately, star spots and random variations in the stars' brightnesses.

A single brief diminution in a star's brightness that has the characteristics of a transit, such as a flat-bottomed minimum, results in the star being labeled as an exoplanetary candidate. If transits alone can be observed then confirmation that an exoplanet has been detected requires the detection of a minimum of three transits (four or five transits would be much better), separated by the same time intervals (i.e. the exoplanet's orbital period). Thus an alien astronomer would need to observe the Sun for a minimum of 24 years (twice Jupiter's orbital period) in order to detect the presence of Jupiter with some certainty. Since it was only in 1999 that the first exoplanet was detected by the transit method, we have over another decade of observations to make before we can hope to find exoplanets in Jupiter-like orbits this way.

Clearly, though, if a Jupiter-sized planet were in a close orbit to its star, the orbital period would be shorter and the discovery could be made that much more quickly. Thus the first exoplanet transit to be observed was that of HD 209458 b (discovered via the radial velocity method) in 1999 and it has an orbital period of just 3.5 days. HD 209458 b is also around 40% larger than Jupiter so that the star's brightness decreases by 1.5% (a dimming by 0.016^m on the normal stellar magnitude scale) during a transit. The planet is just 0.047 AU away from its star giving it a cloud top temperature in excess of 1,000°C. The heating effect of the star has also inflated the planet's atmosphere considerably so that despite being significantly larger than Jupiter its mass is smaller – just 69% that of Jupiter.

In practice the initial observations of transiting exoplanet candidates are often confirmed by separate radial velocity measurements. In any case radial velocity measurements are usually necessary in order to determine all the exoplanet's parameters although the depth of the transit can be used to indicate the size of the exoplanet (Appendix IV) – something that is generally unknown for exoplanets discovered by other methods.

The radial velocity changes of the Sun arising from the Earth's orbital motion are just 100 mm/s, while the current state-of-the-art accuracy in measuring Doppler shifts spectroscopically is still around ten times poorer than that. Until the precision of radial velocity determinations improves by considerably – which is likely to take a decade or more – confirming Earth-sized exoplanets in Earth-type orbits requires a different approach. The transit timing variation method is one such possible approach although it can only be applied to multi-exoplanet systems. In multi-planet systems the planets' gravities pull on each other so that the planets are slightly speeded-up or slowed-down at times compared with their average orbital velocities. This results in the transits occurring very slightly sooner or later than expected. The transit timing variations can then be computer modeled to confirm both the reality of the exoplanets and to give accurate estimates of their masses, orbital periods and orbital sizes. (This is essentially the same method that Urbain Le Verrier and John Couch Adams used in 1846 to predict the position of Neptune in the sky from the changes its gravitational pull produced in the motion of Uranus). The transit timing method has so far been used to confirm the exoplanets in the Kepler-9 (three exoplanets) and Kepler-11 (six exoplanets) systems although the planets involved in both cases are considerably more massive than the Earth and a lot closer to their host stars than is the Earth to the Sun.

HD 209458 b, along with TrES-1, has been observed in eclipse (this should, more properly, be called an occultation but is generally labelled as an eclipse in the literature) as well as in transit. In 2004 NASA's Spitzer infrared space telescope was able to detect the decrease in total radiation from the system by about 0.25% at 24 µm as the planet passed behind its star. At these long infrared wavelengths the planet itself is radiating and the star is relatively dim so that the contrast between them is less overwhelming than that at visual wavelengths.

Like the Doppler approach to exoplanet detection, the transit method is biased towards finding hot Jupiters. The first exoplanets discovered via transits – OGLE-TR-56-b and OGLE-TR-10-b in 2002 – illustrate this well with masses of 0.63 and 1.3 Jupiter masses and orbital periods of 3.1 and 1.2 days respectively. That bias in the case of transiting exoplanets is exacerbated because transits by planets close to their stars are visible over a greater range of angles than those of more distant planets (Figure 6.4). If we take a Jupiter-sized exoplanet and a solar-sized host star, then

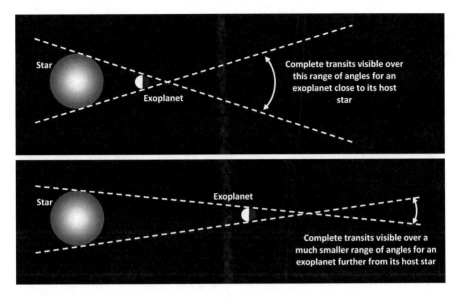

FIGURE 6.4 Transits for the exoplanet closer to its host star are visible from a wider range of angles than is the case for the more distant exoplanet.

4% of such systems will be correctly oriented for us to see transits if the planet and star's separation is 0.1 AU, but only 0.08% if they are as far apart as Jupiter and the Sun (5.2 AU). We are thus more likely to be in the right position to see close exoplanetary transits than those with wider separations.

Observing the transit of an exoplanet is much simpler than determining a Doppler shift and some surprisingly small telescopes – down to 0.1-m – are employed in the task. Measuring brightness also requires equipment that is simple when compared with a sophisticated spectrograph. Not surprisingly therefore, exoplanet transit observation is an area wherein amateur astronomers can and do make a real contribution to research. The problem with transit observation is that in order to pick up a transit in a reasonable amount of time, very large numbers of stars have to be monitored at frequent intervals. Taking the case discussed above of a planet in a 0.1 AU orbit, then 4% or so of such systems will be oriented so that we can see a transit. The transit, however, only lasts for a short while – around 3 h or 4 h every 10 days or so. Thus even if stars could be selected in some way so that they all had exoplanets in 0.1 AU orbits, 1,500 would need to be observed in order to catch one in transit. In reality 10,000 to a 100,000 or more stars need to have their brightnesses measured at intervals of an hour or two over a period of weeks or months in order to succeed in a transit search. Thus the equipment used for transit searches differs from that for simpler photometric tasks mainly by being able to observe many stars simultaneously.

Detection of an exoplanet via its transits gives the orbital period of the planet (from the interval between successive transits), the size of the planet (from the depth of the dip in the star's brightness and the size of the star) and the size of the orbit (from Kepler's third law of planetary motion – Appendix IV). The planet's mass is not determined from the transit information though. Exoplanets discovered via transits are therefore also observed spectroscopically in order to measure their host stars' Doppler shifts. This then fills in the missing information on the exoplanet's mass.

Even with just a single transit, some estimate of the orbital period may be made, though with large uncertainties. It is necessary to assume that the exoplanet passes across the centre of the star's disk during the transit, that the orbit is circular and that the star's radius and mass can be estimated. The duration of the transit then suggests a value for the orbital period (Appendix IV).

Since the planetary radius is not measured for planets with only radial velocity data (only 19% of exoplanets have known radii), we have significantly more information regarding those exoplanets that have been observed both via transits and Doppler shifts (whichever method led to their discovery). Furthermore, not only is the planetary radius known, but the mass that is determined is an actual value, not a minimum. This is because, in order for us to see a transit at all, we know that the exoplanet's orbit must be inclined at very close to 90° to the plane of the sky.

If the host star can be observed spectroscopically during a transit, then it may be possible to measure the angle between the rotation of the star (i.e. its equatorial plane) and the orbital plane of the planet's orbit. With rotating stars, some parts of the star's surface are approaching us. The spectrum lines from those parts of the star are Doppler shifted slightly to shorter wavelengths. Other parts of the star's surface will be moving away from us and the spectrum lines from those parts of the star are shifted to slightly longer wavelengths. When we look at the star's spectrum as a whole these individual Doppler shifts cause the spectrum lines to be slightly wider than they would be if the star were not rotating. At the start of a transit, the planet obscures a small portion of the approaching limb of the star, thus reducing the intensity on the short wavelength edges of the observed spectrum lines. The lines therefore appear to move slightly to longer wavelengths (i.e. a red-shift). At the end of the transit, the planet obscures a small part of the star's receding limb, and the spectrum lines appear to move to shorter wavelengths (a blue-shift). This change in the wavelengths of the star's spectrum lines during a transit of its exoplanet is called the Rossiter-McLaughlin effect and careful computer modeling of the changes gives the angle between the equator and orbital plane. For HD 189733 b, for example, the angle is just 0.85° – as might be expected if star and planet formed from a single rotating nebula (Chap. 13) – for comparison the angle between the Earth's orbit and the Sun's equator is 7.25°.

One highly successful exoplanetary transit hunter with 36 exoplanets on its score sheet (30% of all those discovered from exo-planet transits) is SuperWASP (Super Wide Angle Search for Planets). SuperWASP (a development from a similar but simpler instrument called WASP) comprises two separate instruments. One of these is sited with the Isaac Newton group of telescopes on La Palma in the Canary Islands to cover the northern half of the sky and the other at the South African Astrophysical Observatory in Sutherland to cover the southern half of the sky. Both instruments have eight cameras, each of which images about 61 square degrees. The total sky coverage in a single pointing is thus about 500 square degrees – about 15,000 times larger than the area of sky covered conventional telescopes. Up to a million stars can be imaged in each set of expo-sures (depending upon the star density in the area of the sky being covered) – and exposures are taken every minute throughout the night. Given clear weather therefore some 100 Gbytes of data are obtained every 24 h covering stars down to about 15th magnitude (a 15th magnitude star is about the faintest star that can be seen by eye from a good observing site using a 0.5-m telescope).

SuperWASP's 'telescopes' have apertures of just 0.11 m. They are in fact more-or-less off the shelf telephoto lenses (Cannon™ 200 mm f1.8 lenses) that feed CCD detectors. The cameras are mounted together onto a single mount (Figure 6.5) and the whole

FIGURE 6.5 The SuperWASP North instrument. (Reproduced by kind permission of the SuperWASP project).

instrument operates automatically. The data is also processed automatically, firstly to correct for known problems such as variations in the sensitivity of the pixels and to reduce background noise. The stars are then identified from catalogues and their brightnesses determined. After several months of data have been accumulated the light curves for each star are examined for dips in the brightness that could be due to an exoplanetary transit. Finally stars that have had probable transits detected are observed spectroscopically and the exoplanet discovery (if that is what it is) confirmed by the Doppler shifts of the host star.

The first two SuperWASP exoplanet discoveries, both hot Jupiters, were reported in September 2006 with the confirming spectroscopic observations being made by SOPHIE. WASP-1b in Andromeda is a 0.89 Jupiter mass planet that is 35% larger than Jupiter and which orbits its slightly-larger-than-the-Sun host star at a distance of 0.038 AU every 2.5 days. WASP-2b orbits a star in Delphinus that is slightly cooler, less massive and smaller than the Sun. The exoplanet has a mass of 0.91 Jupiter masses, a radius almost identical to that of Jupiter and an orbital period of 2.15 days.

The 17th SuperWASP exoplanet – WASP-17b, discovered in August 2009 – was a surprise in several respects. It was the first exoplanet to be found whose orbital motion was in the opposite sense to the rotation of its host star. Such retrograde motion is quite unusual but not unknown – within the solar system for example, the well-known Halley's comet has a retrograde orbit. The second surprise came when the exoplanet's size and mass were determined. Its diameter is about 1.7 times that of Jupiter – making it the largest known exoplanet at the time of its discovery – and its mass about 0.5 Jupiter masses. The resulting average density is thus around tenth of that of Jupiter (or about 2% of the Earth's density and about the same density as that ubiquitous lightweight packaging, expanded polystyrene foam). WASP-17b orbits a star somewhat hotter than the Sun every 3.7 days at a distance of 0.05 AU on average. However its orbit is quite elliptical so that the planet's actual distance from the star varies from around 6.5 to 8.5 million kilometres. The changing distance for the planet from its star results in huge tidal stresses inside the planet which heat up its interior. This internal heating combined with the energy coming from the star is sufficient to have led to the enormous bloating of the planet so leading to its extraordinarily low density.

WASP-12 b, discovered in 2008, is in a 0.023 AU orbit around its solar-type host star. This is sufficiently close that material is lost from the planet to the star at a rate of about one Jupiter mass every ten million years. Since the planet's mass is only 1.4 Jupiter masses it seems likely that it will be reduced to its metallic/rocky core in a relatively short time. Spitzer observations have recently shown that WASP-12 b's atmosphere is dominated by carbon compounds, but whether or not this is related to the mass loss is still unclear.

Four searches that are similar to SuperWASP and with successful exoplanet discoveries to their credit are HATNet (Hungarian Automated Telecope Network), TrES (Trans-atlantic Exoplanet Survey), the XO project and the Alsubai project. HAT-Net uses six 0.11-m wide-angle robotic telescopes mainly based on Mauna Kea, Hawaii and at the Smithsonian Astrophysical Observatory in Arizona. Collaboration with a similar instrument based at the Wise observatory in the Negev desert in Israel enhances the sky coverage. Since its first discovery in 2006, HATNet has found 26 exoplanets, mostly hot Jupiters, although HAT-P-11b, discovered in 2009 and with a mass of 0.081 Jupiter masses is only about twice the size of Uranus or Neptune. TrES uses three 0.1-m Schmidt telescopes based at Mount Palomar, the Lowell Observatory in Arizona and the Canary islands. Its four exoplanet discoveries to date are again all hot Jupiters. The XO Project uses two commercial f1.8, 200 mm telephoto lenses on a single mounting and is sited at the summit of Haleakala on Maui, Hawaii. Amateur astronomers as well as professionals are involved in its search. Since 2006, the project has discovered five hot Jupiter exoplanets. The Alsubai project uses a 0.1 m and four 0.035 m cameras and is based in New Mexico. It has recently discovered its first exoplanet, Qatar-1 b, a hot Jupiter orbiting 3.5 million kilometres out from a cool star in Draco.

MEarth (pronounced 'mirth') is a similar system to Super-WASP but using larger telescopes. It comprises eight 0.4-m independently-mounted robotic telescopes housed at the Whipple observatory on Mount Hopkins, Arizona. The project monitors 2,000 small cool stars (red dwarfs) individually for transits. In 2009 a super-Earth was found by the project orbiting a star 40 light years away in Ophiuchus. The discovery was confirmed through radial velocity measurements by HARPS. The star, GJ 1214, is only 0.3% as bright as the Sun and its exoplanet, GJ 1214 b, is about two-and-a-half times the size of the Earth with a mass of six Earth masses (0.018 Jupiter masses). It is the second smallest exoplanet currently known (after CoRoT-7 b – amongst those that have their radii measured). Although the planet is only two million kilometres out from its star, that star is so cool and dim, that the planet is amongst the coolest found so far with a surface temperature of about 200°C. In late 2010 Jacob Bean et al. were able to analyze the atmosphere of GJ 1214 b using VLT observations of the planet obtained during a transit. The near infrared spectrum of the atmosphere turned out to be featureless, ruling out hydrogen as a primary constituent of the atmosphere. The researchers suggest that the atmosphere either has a thick high level cloud layer that masks any hydrogen that may be present or that it contains a high proportion of water vapour (steam).

Yet another robotic planet hunter has recently achieved first light. This is the 0.6-m telescope at La Silla of the TRAPPIST (Transiting Planets and Planetesimals small telescope) project, but it has yet to make any discoveries.

The OGLE (Optical Gravitational Lensing Experiment) project has been operating since 1992 and is led by Prof. Andrzej Udalski of Warsaw University and by the late Prof. Bohdan Paczyński of Princeton University. The details of gravitational lensing are discussed in Chap. 8. Here it is sufficient to note that OGLE can also detect exoplanets via the transit method.

OGLE is now in its fourth phase of development (OGLE IV). OGLE initially used the 1-m Swope telescope at the Las Campanas observatory in Chile. Later, the 1.3-m Warsaw telescope (also at Las Campanas) was purpose built for the project. In the current phase (OGLE IV), the telescope feeds a mosaic of 32 2,048 × 4,096 pixel CCDs giving it a total field of view of 1.4 square degrees. Because the programme's main objective is detecting dark matter, its primary observational targets are the Milky Way's galactic bulge and the Magellanic clouds. For this reason many of OGLE's discoveries are among the most distant known exoplanets. OGLE-TR-56 b, for example, the first exoplanet discovered by OGLE in 2002, is around 5,000 light years away from us. It has a mass of 1.3 Jupiter masses and is in a 29-h orbit just three and third million kilometres away from its solar-type host star. At the time of its discovery OGLE-TR-56 b had the smallest known separation from its host star of any exoplanet – its temperature at the top of its atmosphere is likely to be 1,600–1,700°C, which, since iron melts at 1,538°C, gives rise to the intriguing speculation that there may be clouds of molten iron droplets in the planet's atmosphere and even iron raindrops! OGLE-TR-56 b was amongst a list of over 40 possible exoplanetary transit stars compiled from earlier OGLE observations. Radial velocity measurements by the 10-m Keck telescopes and others showed that most of these transiting objects were too massive to be planets, but OGLE-TR-56 b (and later, in 2004, OGLE-TR-10 b) turned out to be planet-sized. OGLE has now discovered a total of eight exoplanets via the transit approach.

In December 2006 the French CNES (Centre National d'Etudes Spatiales) together with ESA launched the CoRoT (Convection, Rotation and planetary Transits – Figure 6.6) spacecraft with a twofold mission –

1. To study the interiors of stars by observing their vibrations ('stellar seismology' or 'asteroseismology') and
2. To discover Earth-like exoplanets.

FIGURE 6.6 Artist's concept of the CoRoT spacecraft in orbit. (© CNES/ DUCROS David, 2006. Reproduced by kind permission of CNES and David Ducros).

The spacecraft carries a 0.27-m off-axis telescope feeding four 2,048×4,096 pixel CCDs that cover a 2.8°×2.8° area of the sky. Two of the CCDs are devoted to asteroseismology and two to transits, though at the time of writing only one of each is functioning. CoRoT is in 900-kilometre high polar orbit and observes two parts of the sky for 6 months at a time each. The two fields of view are in Aquila and Monoceros and the spacecraft switches between them when the Sun threatens to interfere with the observations in one of the areas. Recently the mission has been extended to continue at least until March 2013. CoRoT's observations are supported by a small ground-based telescope. The Berlin Exoplanet Search Telescope II (BEST II) is a 0.25-m diameter robotic instrument sited near Cerro Armazones in Chile.

CoRoT has discovered and had confirmed 17 exoplanets to date. The first, CoRoT-1 b (Figure 6.7), announced in 2007, is an enormous hot Jupiter orbiting a solar-type star 1,560 light years away from us in Monoceros. The exoplanet has a radius 50% larger than that of Jupiter and a mass equal to that of Jupiter so that its mean density is only a third of that of water.

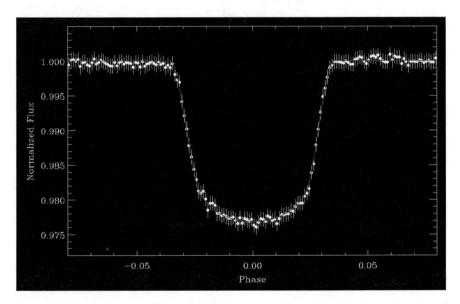

FIGURE 6.7 The transit of CoRoT-1 b. (Reproduced by kind permission of ESA and the CoRoT exo-team).

Most of the CoRoT exoplanet discoveries are hot Jupiters with the exceptions of CoRoT-7 b (2009) and CoRoT-9 b (2010). In CoRoT-7 b (Figure 6.8), the spacecraft's mission to find Earth-sized planets was almost fulfilled. This exoplanet has a radius just 70% larger than that of the Earth (15% of Jupiter's radius) making it the smallest known exoplanet at the time of writing. Its mass is 4.8 times that of the Earth (0.015 Jupiter masses) giving it an average density 5.6 times the density of water (5,600 kg/m^3 compared with 5,500 kg/m^3 for the Earth). It is thus almost certainly of a rocky composition, perhaps with an iron core like the Earth. In other respects though, CoRoT-7 b is not a twin for the Earth. It orbits only 0.017 AU out from its slightly-cooler-than-the-Sun host star so that its surface temperature is variously estimated to be at least 1,000°C and perhaps as much as 2,500°C. The surface is thus likely to be covered by oceans of molten rock and it may have a very thin atmosphere comprised of sodium, oxygen and silicon monoxide. CoRoT-7 b's 'year' is just 20.5 h long – the short-

FIGURE 6.8 An artist's impression of CoRoT-7 b and its host star. (Reproduced by kind permission of ESA, ESO and L. Calcada).

est known for any exoplanet. It is quite likely that the planet's rotation is tidally locked onto its host star so that it always keeps the same face towards the star. The temperature on the side away from the star could then fall as low as –220°C. If the planet's orbit is even slightly elliptical tides could heat up its interior and lead to continuous and intense volcanic activity at the surface. If there is volcanic activity occurring, then the James Webb Space telescope may in due course be able to detect the gases that have been emitted. It is possible that the planet was at one time as large as Neptune and has been evaporated down to its present size (a chthonian planet).

CoRoT-9 b is unusual in that its orbit is relatively large. It is in an orbit larger than that of Mercury (0.41 AU) with an orbital period of 95 days. It is very close to Jupiter in size and has a mass of 0.84 Jupiter masses. Although not a Jupiter-twin, the temperature of its outer layers probably lies between –20°C and 150°C – far cooler than that of the other exoplanets found via the transit approach.

NASA's Kepler mission is based upon a $600 million, 1,000 kg spacecraft that was specifically designed and built with the aim of discovering Earth-sized planets within the habitability zones of Sun-like stars (Figure 6.9). The spacecraft was launched in March 2009 into a Sun-centred orbit (i.e. it does NOT orbit the Earth). The spacecraft's orbital period is 6 days longer than the Earth's year so that it gradually drops further and further behind the Earth at a rate of a million kilometres every 3½ weeks. The orbit was chosen so that the Earth did not block the spacecraft's field of view and so that gravitational disturbances, etc. would be minimized. Sixty-one years after its launch the spacecraft will return to the vicinity of the Earth. There is no possibility of it colliding with the Earth, but what a magnificent opportunity for the salvage experts of 2070!

The spacecraft is built around a large Schmidt camera. The camera has an aperture of 0.95 m and a primary mirror with a diameter of 1.4 m (for comparison the largest ground-based Schmidt camera, at the Karl Schwarzschild observatory, has an aperture of 1.34 m and a primary mirror with a diameter of 2 m). The instrument's field of view encompasses over a 100 square degrees (roughly the area covered by the spread hand held at arm's

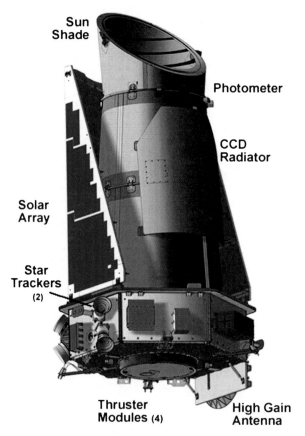

Sun
Shade

Photometer

CCD
Radiator

Solar
Array

Star
Trackers
(2)

Thruster
Modules (4)

High Gain
Antenna

FIGURE 6.9 The Kepler spacecraft. (Reproduced by kind permission of NASA Ames and Ball Aerospace).

length) and its detector is a mosaic of 42 1,024×2,048 pixel CCDs (Figure 6.10). The camera points permanently towards an area of the sky mid-way between Deneb and Vega in the constellations of Cygnus, Lyra and Draco. The area was selected to provide a large number of observable stars, to minimize the number of asteroids and Kuiper belt objects that might be encountered and so that the Sun never interferes with the observations. The volume of space observed by The Kepler spacecraft lies along the Orion spiral arm of the Milky Way, and Earth-sized planets should be detectable out to a distance of 3,000 light years.

A hundred and fifty-five thousand solar-type stars are monitored by The Kepler spacecraft with the CCDs being read out

FIGURE 6.10 The Kepler spacecraft's detectors – the CCD mosaic. (Reproduced by kind permission of NASA Ames and Ball Aerospace).

every 6 s in order to avoid over-exposure. For magnitude 12 stars (a quarter of a million times fainter than Sirius), the stars' brightnesses are measured to a precision of ±0.002%. In every-day terms this level of precision is the equivalent of being able to distinguish the difference in brightnesses between two otherwise identical street lamps, one of which is 100 km (100,000 m) away from the observer and the other which is 1 m closer to the observer (99,999 m). To aid reaching this level of accuracy and perhaps counter-intuitively, the camera is NOT sharply focused. The stellar images are thus a bit fuzzy (about 10 s of arc across) and so are shared amongst 20–30 pixels. An anomalously high or low sensitivity pixel therefore has little effect upon the total measured brightness of the star.

The transit of an Earth-sized planet should produce a drop in the star's brightness by around 0.008–0.009% – about four times larger than the minimum change measureable for a 12^m star. During the mission's scheduled 3.5 year life (which may be extended – the planned life of the spacecraft is 6 years) it is hoped that some 50 Earth-sized planets might be found along with 100 or 200 twice the size of the Earth and up to a 1,000 Jupiter-sized exoplanets. The first exo-Earth discoveries though (Earth-sized and with orbital periods in the region of a year) should not be expected before around 2012–2013 because of the necessity of observing three or more transits. Cold Jupiters (giant planets in long-period orbits) are only likely to have a single transit observed – insufficient to count as a discovery. However the transit for such planets should be readily recognizable as being a transit and will be deep enough to be observed from the ground. Follow-up observations from Earth-based instruments may therefore be used to detect subsequent transits in such cases although the area of sky observed by the Kepler spacecraft is only accessible to such instruments from around about May to October. In this way, Kepler is expected to pin point up to 30 stars that are likely to host cold Jupiter exoplanets. It is also likely that confirming radial velocity observations could be made of these candidate stars without waiting for a second or third occultation and the exoplanets confirmed via that approach.

In the case of Jupiter-sized planets, Kepler should also be able to detect them directly from their reflected light. The exoplanet will change its phase (as Kepler 'sees' it) from zero when it is transiting its host star through a crescent shape, half 'moon', gibbous and finally to full just before it passes behind the star. The phase sequence will repeat in reverse as the planet comes out from behind the star and moves round towards its next transit. This will lead to a small but regular change in the star's brightness that has the same period as that of the exoplanet (Figure 6.11). Since the planet's orbital period will be known very precisely from the transit timings, several of these stellar modulation patterns can be added together to improve their detectability. It is expected that the Kepler spacecraft will be able to observe giant exoplanets in this way when their orbital periods are less than about 7 days. The spacecraft has indeed already observed the effect for the previously-known exoplanet HAT-P-7 b whose orbital period is 2.2 days.

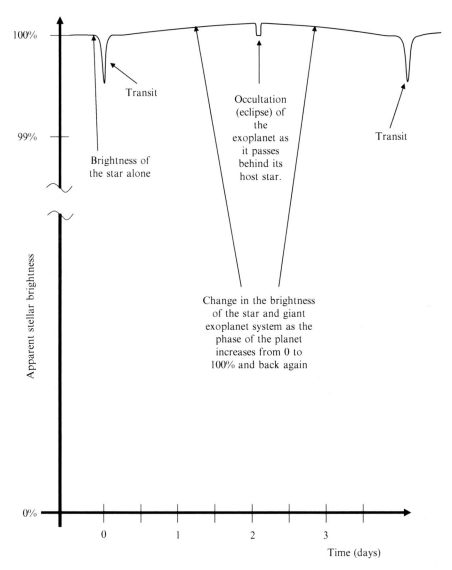

100%

99%

0%

Transit

Occultation
(eclipse) of
the
exoplanet as
it passes
behind its
host star.

Transit

Brightness of
the star alone

Change in the brightness
of the star and giant
exoplanet system as the
phase of the planet
increases from 0 to
100% and back again

Apparent stellar brightness

0   1   2   3

Time (days)

FIGURE 6.11 The modulation of a star and giant exoplanet system's total brightness as the phase of the exoplanet changes.

The Kepler spacecraft commenced observations of its target area of the sky on May 12th 2009. By August of the same year it had confirmed its ability to detect exoplanets by picking up the transit of an already known exoplanet. TrES-2, now also designated Kepler-1 b, was discovered in 2006 by the Trans-atlantic Exoplanet

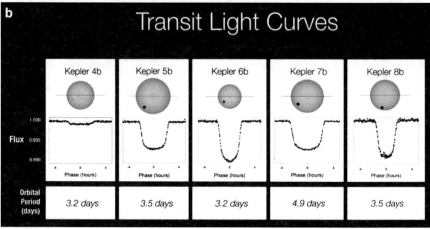

FIGURE 6.12 (a) An artist's impression of the recently discovered pair of hot Jupiters forming the Kepler-9 exoplanetary system. (Reproduced by kind permission of NASA/Ames/JPL-Caltech). (b) Transits of Kepler's first five discoveries. (Reproduced by kind permission of NASA Ames).

Survey and is a 1.2 Jupiter mass hot Jupiter in an orbit with a radius of 0.036 AU around a solar-type star 700 light years away from us on the Cygnus/Draco boundary. A minor problem with the spacecraft developed in November 2009 when 3 of the 84 data channels were found to be noisier than expected. Nonetheless with the January 2010 data release five exoplanet discoveries were announced together with the detections two more previously known planets. All five of the new exoplanets (Figure 6.12b) are hot Jupiters with the exception of Kepler-4 b which is small enough to be classed as a hot Neptune.

In mid-2010 the discovery of three exoplanets transiting the same star in Lyra was announced. Kepler-9 b and Kepler-9 c are Saturn-mass exoplanets in 19 and 38 day orbits 0.09–0.14 AU out from their solar-type host star (Figure 6.12a). While Kepler-9 d is a super-Earth with a mass seven times larger than the Earth. Kepler-9 c is about four million kilometres out from its host star giving it a probable surface temperature around 1,200°C. For the first time the transit timing variation method (Chap. 9) was used, as well as the radial velocity method, to confirm these exoplanets. The discovery of Kepler-10 b was announced in January of 2011 as the first small, rocky planet found by Kepler. It is a super-Earth of 4.6 Earth masses and 40% larger than the Earth. The resulting density of nine times that of water (nearly twice the Earth's average density) is higher than that of iron and suggests that the planet must not only be rocky but have a high proportion of metals such as iron and nickel and a large, highly compressed and dense core.

The February 2011 Kepler data release based upon observations up to mid September 2009 saw the announcement of the discovery of the six-exoplanet system, Kepler-11 (Figure 1.1), with its confirmation being based entirely upon the transit timing variation method. The inner five planets of the system are super-Earths or Hot Neptunes and all are closer to their solar-type host stars than Mercury is to the Sun. Even the outermost member of the system, which possibly has a mass near to that of Jupiter, is closer to the star than Venus is to the Sun.

The 2011 data release also increased the number of exoplanetary candidates to over 1,600 – and the Kepler team expect that some 80% of these will eventually be confirmed to be genuine planets. Of these, 54 were in or near their host stars' habitable zones. While most of the latter were likely to be super-Earths or larger exoplanets, one candidate, KOI 326.01, has a possible size less than that of the Earth and so might be a potential exo-Earth. Statistical analysis of the properties of the exoplanet candidates (which is likely to contain significant uncertainties at the moment) suggests that amongst the solar-type stars selected for study by Kepler we may expect to find that:-

6% of the exoplanets are similar to the Earth in size (less than 1.25 Earth radii)

24% of the exoplanets are super-Earths (1.25–2 Earth radii)

55% of the exoplanets are Neptune-sized (2–6 Earth radii)

14% of the exoplanets are Jupiter-sized (6–15 Earth radii)

1.5% of the exoplanets are larger than Jupiter (15–22 Earth radii) and

17% of the stars with exoplanets have multiple planet systems.

As already noted though, it will take some time, possibly several years, before discoveries of exo-Earth planets can be hoped-for, so this initial detection of a preponderance of large exoplanets close to their host stars is as expected.

A striking aspect of the data on exoplanetary candidates is that the number of candidates peaks at an orbital period of around 2–4 days (an orbital radius of about 0.1 AU). If this is a real effect for genuine exoplanets then it could arise either from the exoplanets' inward migrations coming to a halt close to the star as tides transfer some of the star's rotational energy (angular momentum) to the planets or from the planets breaking-up and crashing into the star – or both effects could be in operation.

Details of the exoplanets discovered by Kepler to February 2011 plus the exoplanet candidate KOI 326.01 (From http://kepler. nasa.gov/) are listed in the following table

Name	Mass (Jupiter masses)	Radius (Jupiter radii)	Orbit radius (AU)	Orbit period (days)	Temperature (top layer of planet's atmosphere or the solid surface) (°C)	Host star mass (solar masses)
Kepler-4 b	0.077	0.357	0.046	3.2	1,400	1.22
Kepler-5 b	2.114	1.431	0.051	3.5	1,600	1.37
Kepler-6 b	0.669	1.323	0.046	3.2	1,200	1.21
Kepler-7 b	0.433	1.478	0.062	4.9	1,250	1.35
Kepler-8 b	0.603	1.419	0.048	3.5	1,500	1.21
Kepler-9 b	0.252	0.842	0.14	19.2	400	1.07
Kepler-9 c	0.171	0.823	0.225	38.9	250	1.07
Kepler-9 d	0.022	0.147	0.0273	1.59	1,200	1.07
Kepler-10 b	0.014	0.127	0.0168	0.837	1,600	0.89
Kepler-11 b	0.0135	0.176	0.091	10.3	600	0.95
Kepler-11 c	0.0425	0.282	0.106	13.0	500	0.95
Kepler-11 d	0.0192	0.307	0.159	22.7	350	0.95
Kepler-11 e	0.0264	0.404	0.194	32.0	300	0.95
Kepler-11 f	0.0072	0.234	0.25	46.7	200	0.95
Kepler-11 g	<0.95	0.327	0.462	118	90	0.95
KOI 326.01	0.002?	0.08?	0.0005?	9?	60?	0.21

As mentioned in connection with the XO project, observing the deeper transits of exoplanets is well within the capabilities of amateur astronomers. There is even a book – *Exoplanet Observing for Amateurs* by Bruce Gary devoted to the topic (Reductionist Publications – first edition available to down-load free of charge from http://brucegary.net/book_EOA/x.htm). Gary was among the first 'amateur' astronomers (though now retired, he spent many years working professionally in the field of planetary radio astronomy) to observe an exoplanet transit – that of HD 209458 b (discovered via the Doppler method in 1999) in 2002 using a 0.25-m telescope. The transit of HD 209458 b had first been observed by non-professional astronomers using a 0.4-m telescope 2 years earlier by a group led by Arto Oksanen working at the Nyrölä Observatory in Finland. Gary's book is full of useful advice on such essentials as flat fielding, obtaining dark frames for the CCD images, limiting the exposures to avoid saturation etc. There is also a useful introductory article – *Imaging Exoplanets* by David Shiga (*Sky and Telescope* magazine page 44, April 2004) that any prospective transit observer is advised to read. Since the main requirement for observing an exoplanet transit is accurate photometry, the tips given in Shiga's article for achieving this are worth summarising (together with a few additions):

1. If you have a choice of observing sites use the photometrically best one – i.e. the one with the least light pollution, the highest altitude (usually), the least likelihood of haze or cloud and the steadiest atmosphere (least scintillation or twinkling of the stars).

2. Try to choose a star and a time of the year so that the star will be at least 45° above the horizon throughout the observing period. Details of known exoplanet transits, predictions of the times of future transits and images for locating the star may be found at the Exoplanet Transit Database (http://var2.astro.cz/ETD/). Predictions of transits are also available at the NStED (http://nsted.ipac.caltech.edu/index.html).

3. Use a comparison star (or several comparison stars) within the same field of view as the transit candidate star which has (have) as similar a colour (spectral type – see Appendix IV for a brief summary of stellar spectral and luminosity classification) and brightness (magnitude) as possible to that of the candidate star.

The widest possible field of view (i.e. the biggest CCD chip that you can afford) will assist in providing suitable comparison stars.

4. Keep the two (or more) stars' images on the same pixels of the CCD camera throughout the observing period.

5. Slightly de-focus the telescope so that the stars' images are spread over several (20–30) pixels. However in crowded star fields be careful that the star images do not start to over-lap.

6. Choose an exposure that is sufficiently short that none of the pixels recording the images of the stars of interest are anywhere near to being saturated. At the same time the exposure should be long enough to even-out variations in brightness due to atmospheric scintillation. In practice this probably means exposures of around 10 s duration. The use of a broad-band filter and/or increasing or decreasing the level of de-focussing of the telescope may help to optimise the exposure. In some CCDs the response starts to become non-linear well before the pixels are saturated. In these cases make sure that the exposures remain within the linear part of the response.

7. Obtain as many images as you can, starting well before the predicted time of transit and continuing until well after its predicted end – you will be unlikely to see the change in brightness whilst still at the telescope.

8. Obtain calibration images (flat field, dark frame, etc.) regularly throughout the observing period – but note point (3) and if you have to move the telescope to obtain the calibrations make sure that the stars images are returned to exactly the same places on the CCD (not easy!!).

9. Keep an accurate record of the times and durations of each exposure together with the usual observing notes regarding weather conditions, instrument problems, observing procedures, etc.

When you have obtained your data, you may be satisfied just with the achievement of detecting the transit. However the sense of achievement will undoubtedly be greater if you can contribute to improving our knowledge of the host star and its exoplanet. To this end there are several programmes that you can join or contribute towards. The American Association of Variable Star Observers (AAVSO – http://www.aavso.org/observing/programs/ccd/transitsearch.shtml), for example, collaborates with Transitsearch (http://www.transitsearch.org/) in observing selected target stars systematically and welcomes contributions from amateur astronomers. AAVSO provides advice and tutorials on how to obtain useful measurements and how to analyse them. A similar scheme, project TRESCA (TRansiting ExoplanetS and CAndidates), is run by the Czech Astronomical Society's Variable Star and Exoplanet section (see Exoplanet Transit Database – http://var2.astro.cz/ETD/).

Amateur astronomers and any other readers with an interest in finding exoplanets can contribute to the analysis of Kepler's data through PlanetHunters.org (http://kepler.nasa.gov/education/planethunters/) – a part of the Citizen Science project (http://citizensciencealliance.org/projects.html). This scheme has over 16,000 contributors at the time of writing, but more are always welcome.

The discovery of a new exoplanet completely through amateur astronomer contributions has yet to occur, although it seems likely to do so fairly soon. Amateur astronomers though have contributed to exoplanet discoveries, working in collaboration with professionally-operated searches. The five XO project planet discoveries have already been mentioned as one such example. Another is for the exoplanet HD 17156 b that orbits a solar-type star 250 light years away from us in Cassiopeia. HD 17156 b was discovered in April 2007 through the radial velocity method by a team led by Debra Fischer using observations from the 10-m Keck and 8.2-m Subaru telescopes. The possibility of the orbit being suitably aligned with the Earth for transits to occur seemed likely and in October 2007 a transit was indeed observed by several groups including members of Transitsearch. HD 17156 b turns out to be in a highly elliptical orbit, at one point being 0.27 AU from its host star but then approaching to within 0.05 AU of the star at the opposite point of its orbit. Similarly the transit of HD 80606 b (discovered through radial velocity variations by Mayor and Queloz et al. in 2001) was detected by Transitsearch participants amongst others in 2009. Somewhat remarkably, HD 80606 b is in an even more elliptical orbit than HD 17156 b, ranging from 0.03 to 0.88 AU away from its slightly-cooler-than-the-Sun host star. The orbital eccentricity of 0.93 for HD 80606 b is the second largest of any known exoplanet and comparable with that of the orbit of Halley's comet. Amateur astronomers have also made confirming and follow-up observations for many other transiting exoplanets.

7. On the Track of Alien Planets – Direct Imaging and Observation (~2.9% of All Exoplanet Primary Discoveries or ~6% if Free Floating Planets are Included)

Unlike the Doppler and transit approaches to discovering exoplanets with their inbuilt biases towards discovering hot Jupiters, finding exoplanets by direct observation is biased towards detecting exoplanets that are a long way out from their host stars. Thus the orbital radii of those directly observed exoplanets currently known ranges from at least 4 to nearly 700 AU – several thousand times further from their host stars than most of the exoplanets considered up to now. The reason for this bias is obvious – the light from the much brighter star swamps the light from the exoplanet unless they are well separated from each other.

An ET astronomer trying to image Jupiter from a distance of 10 light years would find the Sun 400,000,000 times brighter than the planet and at maximum separation they would be just one and a half seconds of arc apart. If the ET astronomer were a 1,000 light years away, the Sun and Jupiter would still have the same relative brightnesses, but their separation would be just fifteen thousandths of a second of arc. Thus there is a second bias involved in directly imaging exoplanets – it is much easier to see the planets around nearby stars than those around more distant stars.

It is even easier to see planets when they have no host star at all. Direct observations of free-floating planets were thus obtained in 1998 – 6 years before the first direct image of an exoplanet

C. Kitchin, *Exoplanets: Finding, Exploring, and Understanding*
Alien Worlds, Astronomers' Universe, DOI 10.1007/978-1-4614-0644-0_7,
© Springer Science+Business Media, LLC 2012

belonging to a star. Of course, without a host star to provide illumination, free-floating planets cannot be seen by reflected light. They are therefore detected by their own emissions in the infrared and microwave regions of the spectrum.

The first free-floating planets were found as part of a search for brown dwarfs by Motohide Tamura *et al* within the Taurus and Chamaeleon molecular clouds. Using the 2.2-m and 1.5-m telescopes of the University of Hawaii and the Cerro Tololo inter-American observatory and observing in the near infrared, the team found numerous 'very low luminosity young stellar objects' which they interpreted to be brown dwarfs. Even fainter than these objects, they found some 'extremely low luminosity young stellar objects' and within the Chamaeleon cloud they picked up several of these that were '… apparently isolated single sources'. By assuming an age for the objects of a few million years, it is possible to estimate their masses. For five of these single sources, the masses came out to be less than or equal to around 12 Jupiter masses – at the upper end of the mass range for planets but below that required for brown dwarfs.

Phil Lucas and Pat Roche used the 3.8-m UK Infrared tele-scope (UKIRT) 2 years later to observe numerous very young brown dwarfs within the Trapezium cluster of the Orion nebula. Along with the brown dwarfs, they discovered 13 objects with masses less than 13 Jupiter masses and one with a mass of eight Jupiter masses (Figure 7.1). No objects with masses lower than 8 Jupiter masses were detected – perhaps, as suggested by Lucas and Roche, because they did not have time to form before the gas cloud was dispersed by very hot stars. Several similar isolated super Jupiters were found at about the same time in the σ Ori cluster by Rafael Rebolo et al. In 2008, Alexander Schultz and Ray Jayawardhana used the Spitzer spacecraft to detect an isolated three Jupiter mass object, S Ori 70, in the σ Orionis cluster. While in 2010 Kenneth Marsh and his colleagues found a two to three Jupiter mass free floating planet in the ρ Oph cloud from near infrared observations obtained with the 10-m Keck I telescope. The temperature of the outermost layer of that exoplanet is about 1,100°C.

We are able to observe these free-floating planets because they are still relatively warm from the residual heat left over from their formation. Older free-floating planets will have

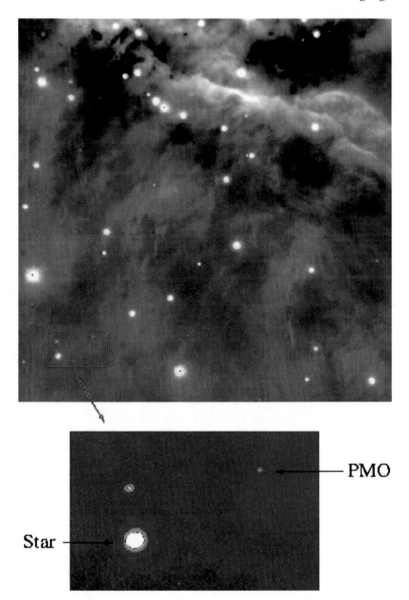

FIGURE 7.1 Ori-188-658, a free floating planet (planetary-mass object – PMO) in Orion. This is a false-colour image in the near infrared obtained in 2001 using the Flamingos 1 camera on the 8.1-m Gemini South telescope. The bright nebulosity at the top of the top image is a part of the Orion bar and the Trapezium would lie just a small distance further up, but is outside the image. (Reproduced by kind permission of P. Lucas).

cooled down towards the temperature of inter-stellar space and so be unobservable using current techniques. Clearly, though, free floating planets are present in abundance in at least a few star-forming regions and we should expect them to be present throughout the galaxy. Because cold free-floating exoplanets cannot be observed, but may be expected to be a 100 or more times commoner than the hot ones, a reliable estimate of the number of such planets is currently unobtainable. However two lines of evidence suggest the number could be very large. Firstly, the lower mass brown dwarfs seem to be commoner than those of higher mass and if that trend continues, it would imply the existence of an even larger number of free floating planets. Secondly, gravitational microlensing observations of the globular cluster, M 22, by Kailash Sahu *et al*, have found events that could be due to free-floating planets occurring more frequently than those due to stars. It thus seems likely that the number of free floating planets at least equals the number of visible stars and that they quite possibly could be thousands of times more numerous.

Though it may seem unlikely, older free floating planets might still be habitable. Energy leaking out from their hot interiors, if trapped by a very thick atmosphere, could raise the temperature at the solid surface (if any) of an Earth-sized object to the point where liquid water might exist and so some form of life (Chap. 14) might be possible.

The first 'normal' exoplanet to be directly imaged was discovered in April 2004 by Gaël Chauvin *et al*. However the exoplanet was not orbiting a star, but a brown dwarf 170 light years away from us in the TW Hydrae association. The brown dwarf, known as 2MASSW J1207334-393254 (usually shortened to 2M1207) actually lies within the constellation of Centaurus and is only about 0.2% as bright as the Sun so is just a 100 times brighter than its planet (Figure 3.14).

The comparatively low contrast between the host object and planet plus a separation of 0.8 s of arc made discovery of the planet, 2M1207 b, relatively easy. However, in this context, 'relatively easy' means using state-of-the-art equipment and the largest of telescopes. Chauvin and his team thus employed the NACO imager on the 8.2-m Yepun telescope of ESO's VLT. NACO (NAOS-CONICA) combines an adaptive optics system (NAOS – Nasmyth

Adaptive Optics System) with a near infrared coronagraphic imager (CONICA – Coudé Near Infrared Camera). The adaptive optics system (Appendix IV) corrects the distortions caused by the Earth's atmosphere enabling the telescope to work at very close to its theoretical resolution (0.04 s of arc for a wavelength of 1.2 μm). The coronagraphic imager (Appendix IV) reduces the contrast between the two objects until the fainter one is no longer swamped by the brighter. The fainter object was confirmed in early 2005 to be connected to the brown dwarf (and hence not a background object) by further VLT and by HST observations. 2M1207 b is now known to have a minimum mass four times that of Jupiter and to be in a ~ 2,500-year orbit about 50 AU out from the brown dwarf. The planet's cloud top temperature is about 1,300°C and its radius about 1.5 Jupiter radii. The exoplanet's spectrum shows signs of water vapour in its atmosphere. For comparison the brown dwarf host has a mass of 25 Jupiter masses, a radius of 2.5 Jupiter radii (0.25 solar radii) and a surface temperature of 2,300°C.

A direct image of a faint object close to a cool but normal star some 150 light years away from us in Pictor was obtained a year earlier than that of 2M1207 b, also using the NACO instrument on the VLT. AB Pic b is about 5 s of arc away from the star (Figure 7.2a), corresponding to a physical separation of about 275 AU and has a surface temperature of around 1,800°C. However current estimates of its mass suggest that it is at least 13–14 Jupiter masses – putting it into just into the brown dwarfs rather making it a very large exoplanet. Also ambiguous between being an exoplanet or brown dwarf, though perhaps with more chance of the former, is GQ Lup b (Figure 7.2b). This object was imaged in 2005, yet again with the ESO's NACO/VLT combination. The star is around 500 light years distant and its companion is 0.7 s of arc away from it – corresponding to a physical separation of 100 AU. Its mass is very uncertain – probably lying between 1 and 36 Jupiter masses.

Several other low mass stellar companions were imaged during the next couple of years, but all seemed more likely to be brown dwarfs than planets. The first companion to a normal star that had a reasonable probability of being an exoplanet was thus not seen until 2008. Then, in a period of just 2 months, six directly imaged and almost certain exoplanets belonging to normal stars, including a planetary system with three exoplanets, were announced (Figure 7.3).

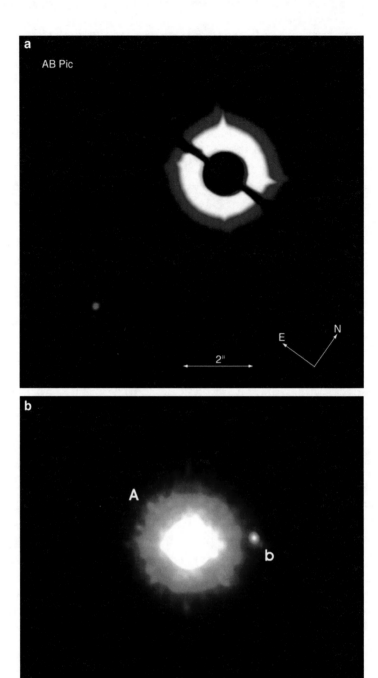

Figure 7.2 (a) A NACO/VLT coronagraphic image of AB Pic (the bright object just above centre) and its brown dwarf/exoplanet companion. The shadow of the coronagraphic mask (Appendix IV) and its supports can be seen silhouetted against the star's scattered light. (b) A NACO/VLT coronagraphic image of GQ Lup and its companion. (Reproduced by courtesy of ESO).

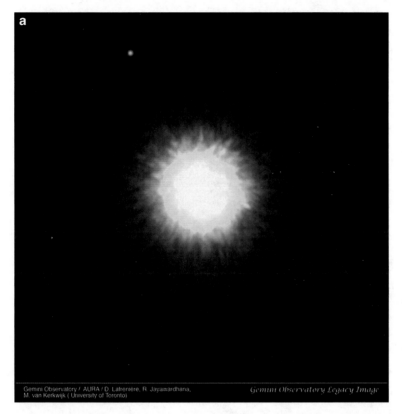

Gemini Observatory / AURA / D. Lafrenière, R. Jayawardhana,
M. van Kerkwijk (University of Toronto)   *Gemini Observatory Legacy Image*

FIGURE 7.3 (a) Gemini image of 1RXS J160929.1-210524 and its possible exoplanet. The exoplanet is the faint object towards the top left of the image (Reproduced by kind permission of Gemini Observatory, AURA, D. Lafrenière, R. Jayawardhana and M. Van Kerkwijk (University of Toronto)). (b) Fomalhaut and exoplanet (Reproduced by kind permission of Paul Kalas/NASA/ESA). (c) HR 8799 and its exoplanetary system (Reproduced by kind permission of Gemini Observatory, NRC, AURA and Christian Marois et al.). (d) A composite image showing the disk of β Pic and its exoplanet. The disk is shown from an infrared image obtained with ESO's 3.6-m telescope in 1996 and combined with the 2003 and 2009 corona-graphic NACO/VLT 3.6 µm images showing β Pic b. (Reproduced by kind permission of ESO and A.-M. Lagrange et al.).

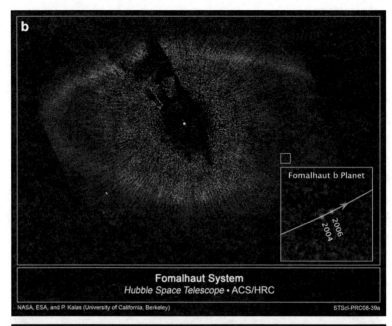

Fomalhaut System
Hubble Space Telescope • ACS/HRC

NASA, ESA, and P. Kalas (University of California, Berkeley)   STScI-PRC08-39a

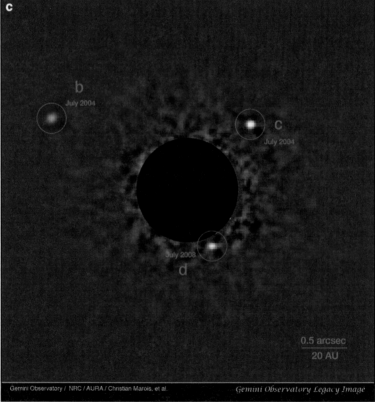

FIGURE 7.3 (continued)

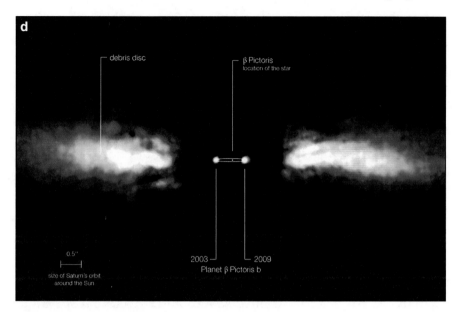

d

debris disc

β Pictoris
location of the star

0.5"
size of Saturn's orbit
around the Sun

2003 ⌐ ⌐ 2009
Planet β Pictoris b

FIGURE 7.3 (continued)

On the 22nd September 2008, David Lafrenière *et al* announced that they had imaged a faint object about 2.5 arc sec distant from a star some 470 light years away from us in Scorpio using the 8.1-m Gemini North telescope operating with adaptive optics. The star, 1RXS J160929.1-210524 (abbreviated to 1RXS 1609 normally), is a young T Tauri variable star that may be expected to settle down to become a normal smallish main sequence star in the next few million years. The companion, 1RXS 1609 b (Figure 7.3a), which was confirmed as being gravitationally linked to the star in 2010, is 330 AU out from the star. Its mass, estimated from its supposed age of around 5 million years and a spectroscopically determined temperature of 1,500°C, is ~8 Jupiter masses.

November the 13th 2008 was a red-letter day for imaging exo-planets – it saw the announcement of a directly imaged exoplanet belonging to the first magnitude southern star, Fomalhaut (also known as α PsA), whilst simultaneously the announcement was made of a directly imaged planetary system comprising three exo-planets for the star HR 8799.

Fomalhaut is a hot star at a distance of just 25 light years and is embedded within a dust belt. The dust belt has a sharp inner boundary and is off-set from the star, so the presence of an exoplanet had been suspected as early as 2005. Fomalhaut b was first imaged by the HST at visible wavelengths using the coronagraphic mode of the Advanced Camera for Surveys in 2004 and then again in 2006 (Figure 7.3b), but it was not until 2 years later that Paul Kalas and James Graham were able to confirm that it was orbiting the star. The exoplanet is in an 872-year orbit that is slightly elliptical so that it ranges from 103 to 128 AU away from its host star. Fomalhaut b's physical size is similar to that of Jupiter, its mass is certainly less than three Jupiter masses and probably lies between 0.5 and 2 Jupiter masses, its temperature may be as low as 100°C and it may have rings of ice and dust similar to, but several times larger than, those of Saturn.

HR 8799 is a sixth magnitude star (and therefore just visible to the naked eye from a good site) in Pegasus. It is somewhat hotter, brighter and more massive than the Sun, surrounded by a huge disk of gas and dust and lies at a distance of 130 light years. Christian Marois and his team observed it in the infrared with both the 10-m Keck and 8.1-m Gemini telescopes from 2004 onwards, in both cases sharpening the images with adaptive optics systems. The orbital motions of the exoplanets (anti-clockwise as seen from the Earth) have been observed and confirm that they are planets belonging to HR 8799 and are not background objects. The innermost exoplanet HR 8799 d was not found until 2008 and so has a 'later' letter than the two outer planets, HR 8799 b, and HR 8799 c. The physical details of the three planets are listed below.

Parameters of HR 8799's exoplanets

Planet	Mass (Jupiter masses)	Radius (Jupiter radii)	Temperature (°C)	Orbital radius (AU)	Orbital period (Years)
HR 8799 d	~10	1.2	?	24	100
HR 8799 c	~10	1.3	~800	38	190
HR 8799 b	~7	1.1	?	68	460

HR 8799 c had been imaged by the HST's NICMOS instrument (Near Infrared Camera and Multi-Object Spectrometer) as early as 1998, but it was only picked up in 2009 from archive data when improved image processing techniques were developed. Perhaps the most remarkable observation of the HR 8799 system though came in January 2010 when Markus Janson *et al* using the VLT and NACO obtained a direct infrared spectrum of HR 8799 c (Figure 7.4). The spectrum shows differences from that expected theoretically which may arise through the presence of dust or clouds in the planet's atmosphere.

Finally on the 21st November 2008, a direct image of a probable exoplanet belonging to β Pic was announced by Anne-Marie Lagrange and others. The nearby naked-eye star β Pic is well-known for being surrounded by a thick disk of gas and dust initially detected by the IRAS spacecraft in 1983 (Figure 3.12). Structures within the central parts of the disk, such as belts and rings of material and dust-free gaps, had suggested the presence of one or more exoplanets for some time. NACO/VLT images of the system were obtained in 2003 to look for such a planet. However it was not

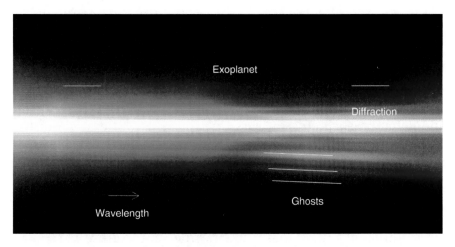

FIGURE 7.4 The spectrum of HR 8799 and one of its exoplanets – HR 8799 c. NB Despite the rather misleading visual colours that are shown here the spectrum is actually in the near infrared at around 4 μm wavelength. (Reproduced by kind permission of ESO and M. Janson).

until 2008, when that data was re-analyzed, that the possible exoplanet was identified. Confirmation that it is an exoplanet and not a background or foreground object came in June 2010 when after disappearing behind or in front of the star in 2008 and 2009, the planet reappeared on the other side of the star (Figure 7.3d). The planet's mass is estimated at around eight Jupiter masses and it is in a ~17 to ~35-year orbit 8–15 AU out from its host star.

Since the 'Super-Thursday' of November 13th 2008 and up to the time of writing, only one more probable exoplanet direct image has been obtained although a couple of smallish brown dwarfs have also been detected. The exoplanet belongs to a small (20 Jupiter masses) brown dwarf 450 light years away from us in Taurus labeled 2M J044144. The planet, 2M J044144 b, has a probable mass of ~7 Jupiter masses and is in a 15 AU, ~400 year orbit. A small star and another brown dwarf may be gravitationally linked to the brown dwarf making the system a quadruple one overall.

Although not directly imaged in quite the same way as the exoplanets just considered, in 2007 Heather Knutson *et al* used the Spitzer spacecraft for 33 h to observe HD 189733 b and to produce the first ever map of the surface of an exoplanet (Figure 7.5). The host star, HD 189733, is to be found in Vulpecula and is about 63 light years away from us. The star's exoplanet was discovered in 2005 by the Doppler method and is very similar to Jupiter (1.13 Jupiter masses, 1.14 Jupiter radii) but orbits its star in just 2.2 days at a distance of 0.03 AU. Knutson and her team observed the planet at a wavelength of 8 μm as it orbited its star and detected an increase in its brightness as the dayside of the planet rotated into view. The hottest part of the planet (940°C) does not face directly towards the star, but is displaced by about 30° of longitude towards the East. This is probably due to winds of up to nearly 10,000 km/h rushing round towards the planet's coolest spot (700°C) on its side furthest from the star. The data were compiled into a map showing the temperature variations over the cloud tops of the planet at a resolution of about a quarter of the planet's radius (about 1.5 times the size of the Earth). More recently Spitzer has identified a hot spot on υ And b, a hot Jupiter

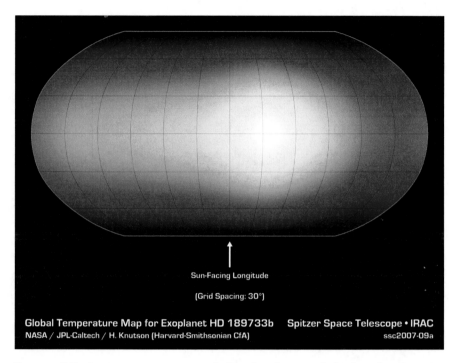

Global Temperature Map for Exoplanet HD 189733b   Spitzer Space Telescope • IRAC
NASA / JPL-Caltech / H. Knutson (Harvard-Smithsonian CfA)   ssc2007-09a

FIGURE 7.5 A temperature map of the cloud tops of the hot Jupiter exoplanet, HD 189733 b. NB The arrow marking the 'Sun-Facing Longitude' actually marks the sub-stellar point on the planet, not the direction towards ourselves. (Reproduced by kind permission of NASA, JPL-Calthech and H. Knutson (Harvard-Smithsonian CfA)).

orbiting nine million kilometres out from a star that is a bit warmer than the Sun. In this case the hot spot is 80° round from the sub-stellar point – almost into the dark side of the planet. The reasons for such a large offset are not clear, although shock waves or magnetic fields may be involved.

8. On the Track of Alien Planets – Gravitational Microlensing (~2.3% of All Exoplanet Primary Discoveries)

The detection of exoplanets via gravitational microlensing is closely related to the transit detection method in that both approaches require regular, very precise photometry of many stars over periods of years. Microlensing though, produces a brief apparent one-off increase in the star's brightness and the physical mechanism involved is quite different from that for a transit.

Details of how gravitational lensing and gravitational microlensing enable exoplanets to be detected are to be found in Appendix IV. Here only a brief outline is given of the process for quick reference.

Imagine first a large flat screen placed perpendicular to the light coming from a distant star. The screen would have a faint uniform illumination from that star. However, the gravitational field of a star acts like a very poor quality and very weak lens. If a second star were thus to be interposed between the first star and the screen, the second star's gravitational field would bend the paths of the light rays very slightly and the illumination of the screen would no longer be uniform. Now imagine that the nearer star hosts an exoplanet. The gravitational field of the exoplanet will also bend the light from the more distant star producing a ripple pattern in the screen's illumination not too dissimilar to that seen on the bottom of a swimming pool. Since both of the stars and the exoplanet will be moving, the ripple pattern will change over a few days as their mutual alignment changes.

C. Kitchin, *Exoplanets: Finding, Exploring, and Understanding Alien Worlds*, Astronomers' Universe, DOI 10.1007/978-1-4614-0644-0_8, © Springer Science+Business Media, LLC 2012

Now replace the screen with an exoplanet hunting team and their telescope. The telescope will only pick up the light from a very small part of the screen at any given moment, but the observed intensity of the light will alter as the ripples move across the telescope. If there is no exoplanet, then the brightness will vary with a smooth rise and fall over a period ranging from a day or so to a month or two. If the nearer star is hosting an exoplanet however, then that smooth variation in intensity will have a number of spikes (sudden brightness increases) and sharp decreases in brightness superimposed upon it. It is this alteration in intensity that enables the exoplanet to be found.

The motive behind attempts to observe microlensing events was initially to try to detect some of the 'missing mass' of the universe. It was thought that some or all of the missing mass might be in the form of objects such as black holes, neutron stars, brown dwarfs, white dwarfs, planets, small stars, etc. that could not be observed directly. In 1986 Bohdan Paczyński suggested that if such an object were to pass in front of a more distant star its gravitational field would lens the light from that star and the event might be detectable. This led, 10 years later, to him and others setting up the OGLE experiment to search for microlensing events.

In the mean time Paczyński and Shude Mao also proposed that exoplanets could be discovered through their effect upon the light curve of a microlensing event. Writing in 1991 they prophesied "A massive search for microlensing of the galactic bulge stars may lead to the discovery of the first extrasolar planetary system." (Astrophysical Journal **374**, L37). In fact, as we have seen, the first exoplanet discoveries came through other approaches, however in 2003 a microlensing event did lead to the discovery of a 2.6 Jupiter mass exoplanet five astronomical units out from a small star some 17,000 light years away from us towards the galactic centre. Since then another eleven exoplanets including two orbiting a single star have been discovered through microlensing events.

Although there is some bias with microlensing towards finding the more massive exoplanets, this is much less so than with the previously discussed approaches. Microlensing can in fact detect low mass planets relatively easily and MOA-2007-BLG-192-L b at 0.01 Jupiter masses (three Earth masses) is the second smallest exoplanet currently known orbiting a normal star. The bias for exoplanets discovered through microlensing events is for them to be at large distances away from us since this increases the probability of two stars (the lensing star and the lensed star) being close to the same line of sight. Also since microlens searches deliberately target regions of the sky such as the centre (bulge) of our own galaxy, the Andromeda galaxy (M31) and the large Magellanic cloud (LMC) where there are many stars, there is a bias towards finding exoplanets from such regions – indeed microlensing is the only current exoplanet search method with the potential for discovering planets outside the Milky Way galaxy, although it has yet to do so. Finally there is a bias towards finding exoplanets separated from their stars by the Einstein radius (Appendix IV) since this produces the largest deviations of the light curve.

Clearly microlensing events are unique. The chances of two stars being sufficiently well aligned for an event to occur is about one in a million for stars towards the galactic centre and one in ten million for stars towards the LMC. The probability of a lensing star aligning with a second source star must therefore be around one in a million million. Furthermore the lensing star will need to travel several seconds of arc or more across the sky to align with the next source star and since the rate of motion usually measured in milli-arc seconds per year this will require an elapsed time ranging from centuries to many thousands of years before the repeat event can occur. On the other hand once two stars are sufficiently closely aligned for a microlensing event to occur the probability of the existence of a potentially observable exoplanet near the lensing star may be as high as 20%.

The 2003 exoplanet lensing discovery was made by the OGLE and MOA searches independently. The OGLE search has been described earlier. MOA (Microlensing observations in Astrophysics) initially used the 0.6-m telescope at the Mount St John observatory in New Zealand's South island and now uses a purpose-built 1.8-m instrument. It is a collaboration between Japanese and New Zealand scientists led by Yasushi Muraki. The OGLE team announced the start of a microlensing event on 22nd June 2003 and the MOA team picked it up a month later. The event was thus labeled OGLE 2003-BLG-235 and MOA 2003-BLG-53. It lasted for about 80 days with a dramatic deviation from the smooth variation for a pair of single stars from the 14th to the 21st July (Figure 8.1). The separation of the exoplanet from its host star

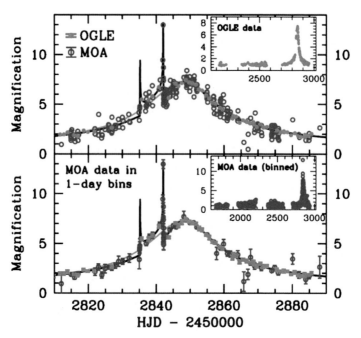

FIGURE 8.1 The microlensing event of 2003 observed by OGLE and MOA that resulted in the discovery of OGLE-235-MOA-53 b. The solid line gives the best-fit computer model of the event and was arrived at by three different groups using different methods. (Astrophysical Journal, **606**, L155, 2004, I.A. Bond et al – Reproduced by kind permission of the AAS and I. A Bond).

was 1.12 times the Einstein radius at the time of the microlensing event. The observations suggest that the lens and source are moving with respect to each other at about three milli arc-seconds per year so by about the year 2013 it should be possible to make confirming observations of the stars individually using a large ground-based adaptive optics telescope or with the JWST.

Not all microlensing events due to exoplanets are as obvious as that shown in Figure 8.1. Figure 8.2 shows the light curve of MOA-2008-BLG-310. Only when the observed curve is carefully compared with the theoretical curve for both the lens and source being single stars do the deviations induced by the exoplanet become apparent. MOA-2008-BLG-310 b has a probable mass of ≥0.23 Jupiter masses and a separation from its host star of 1.25 AU. It is probably within the galaxy's central bulge and at a distance from Earth in excess of 20,000 light years.

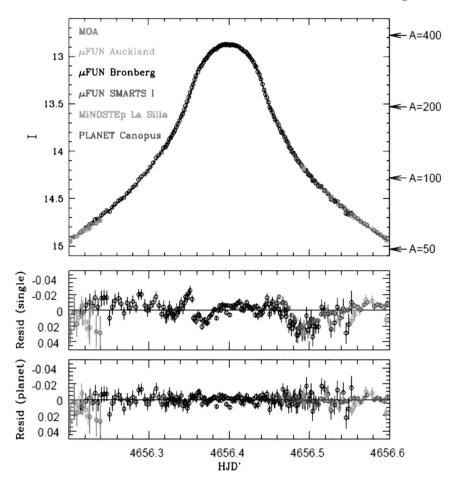

FIGURE 8.2 The microlensing event on 8th July 2008 for MOA-2008-BLG-310. The top curve shows the gross light curve resulting from the combined observations of six groups. The middle plot shows the deviations between the observed light curve and the theoretical one for two single stars. Deviations can clearly be seen at HJD′ (Heliocentric Julian date minus 2,450,000) = 4656.34 and HJD′ = 4656.48 (8 pm and 11 pm UTC). The final curve shows the deviations of the observed curve from the theoretical model for a lensing object comprising a star and an exoplanet. (Astrophysical Journal, ApJ **711**, 731, 2010, J.Janczak et al – Reproduced by kind permission of the AAS and J. Janczak).

In one case microlensing has revealed a planetary system comprising 0.73 and 0.27 Jupiter mass exoplanets. The OGLE-2006-BLG-109 microlensing event lasted for about 20 days in March and April 2006. Early deviations from the single star light curve led to a prediction that the presence of a Jovian class exoplanet should cause a spike in the light curve on the 8th April. This spike was indeed observed, but so also was an unexpected one on the 5th and 6th April. This latter spike arose from the presence of the second exoplanet. OGLE-2006-BLG-109 L b and OGLE-2006-BLG-109 L c are separated from their low mass host star by 2.3 and 4.6 AU respectively and the system is around 5,000 light years away from us in the direction towards the galactic centre.

The details of all exoplanets discovered through microlensing events at the time of writing are listed below. These planets are all to be found in the sky in the direction towards the centre of the Milky Way galaxy in Sagittarius.

Parameters of microlens-discovered exoplanets – in order of their discovery (http://exoplanet.eu/).

Planet	Mass (Jupiter masses)	Separation from star (astronomical units)	(milli arc-seconds)	Orbital period (years)	Distance (light years)
OGLE235-MOA53 b	~2.6	~5.1	~1		17,000
OGLE-2005-071L b	3.5	~3.6	~1	~10	~11,000
OGLE-2005-169L b	0.04	2.8	1	9	9,000
OGLE-2005-390L b	0.017	2.1	0.3	10	21,000
MOA-2007-BLG-192L b	0.01	~0.6	~0.7		~3,000
MOA-2007-BLG-400L b	~0.9	~0.8	~0.1		20,000
OGLE-2006-109L b	0.73	~2.3	~1.5	~5	5,000
OGLE-2006-109L c	0.27	~4.6	~3	~14	5,000
OGLE-2007-BLG-368L b	0.07	~3	~<0.5		>19,000
MOA-2008-BLG-310L b	0.23	1.25	<0.2		>20,000
MOA-2009-BLG-319 b	0.157	~2	~0.3		~20,000
MOA-2009-BLG-3871 b	2.6	~1.8	~0.3		~20,000

OGLE and MOA are primarily monitoring programmes and one of their major functions is to alert other observers whenever a microlensing event is identified. OGLE typically issues 500 microlensing event alerts per year while MOA issues around 50. Both OGLE and MOA can switch from survey to follow-up mode for events of particular interest. Whilst OGLE and/or MOA data usually form a part of the observational campaign of a microlensing event, numerous other groups and observatories will also be involved.

Amongst the many follow-up groups at one time were MPS (Microlensing Planet Search), GMAN (Global Microlensing Alert Network), MACHO (Massive Compact Halo Object) and Super-MACHO surveys, PLANET (Probing Lensing Anomalies NETwork) and Robonet. Currently MicroFUN (Microlensing Follow-Up Network - http://www.astronomy.ohio-state.edu/~microfun/) is a large consortium including many amateur astronomers that is spread over five continents and which concentrates on observing high magnification microlensing events (Appendix IV) in detail. The majority of the telescopes used by the consortium members are relatively small – 0.25 to 0.4 m in diameter - though there are some metre and 2-m class instruments involved as well. PLANET and Robonet were originally independent follow-up programmes using 1- and 2-m class telescopes. Since January 2009 PLANET and Robonet have merged with MicroFUN to form a single group; the MicroFUN-PLANET collaboration (http://planet.iap.fr/).

9. On the Track of Alien Planets – Timing (~1.9% of All Exoplanet Primary Discoveries)

The potential to discover exoplanets by the disturbances that they cause in repetitive phenomena arises with four groups of objects – exoplanets having pulsars (neutron stars) as their host stars, exoplanets having regularly pulsating stars as hosts, exoplanets with eclipsing binary stars as their hosts and transiting exoplanets.

Four exoplanets are known from the first group – PSR 1257+12 B, C and A and PSR B1620-26 b. A fifth companion to a pulsar stretches even the liberal definition of an exoplanet adopted for this book. The milli-second pulsar, XTE J0929-314 was discovered in 2002 using the Rossi X-Ray Timing Explorer spacecraft. It is accreting mass from a companion that is in a 43-minute, 0.002 AU orbit and which has a mass around ten Jupiter masses. In this case however, the companion was almost certainly once a star with a mass 50% of that of the Sun or more which had evolved to a white dwarf. The loss of mass to the neutron star which is the pulsar has reduced the companion's once stellar mass to that of a large exoplanet[1]. Since we have defined planets as objects that do not nor ever have supported fusion reactions, we will not consider these objects any further. PSR 1257+12 B and C were the first exoplanets to be found and along with PSR 1257+12 D and PSR B1620-26 b their method of detection has been discussed earlier. Here, therefore, we shall look at the discoveries of exoplanets via stellar pulsations, binary star eclipses and transits.

The eclipse of an eclipsing binary star is just a larger scale version of the transit of an exoplanet. Two stars are gravitationally bound to each other and the plane of their orbit lies close to the line

C. Kitchin, *Exoplanets: Finding, Exploring, and Understanding*
Alien Worlds, Astronomers' Universe, DOI 10.1007/978-1-4614-0644-0_9,
© Springer Science+Business Media, LLC 2012

of sight from the Earth. Twice every orbit therefore one star wholly or partially passes behind the other. Since both stars are luminous objects their combined brightness falls when a part of one of the stars' surfaces is obscured by the other. The regular fading of an eclipsing binary star was first noted for Algol (β Per) in 1782 by John Goodricke (although he thought that the obscuring body was dark). In fact the fading, but not its regular occurrence, had been spotted over a century earlier and the name Algol itself (from the Arabic *al-ghūl* meaning ghoul or demon) suggests that its anomalous behaviour has been known for millennia. The eclipse of the star with the higher surface brightness per unit area (the hotter star) produces a deeper fading than the eclipse of the other star. The deeper eclipse is then called the primary eclipse and the shallower one, the secondary eclipse. In the case of Algol the temperatures of the two stars (Algol A and Algol B) are so different (11,700–4,200°C) that the primary eclipse is 15 times deeper than the secondary eclipse.

The primary eclipses of Algol occur at intervals of 2 days 20 h 48 min 57 s and the two stars are separated by 0.06 AU. However there is a third star in the system at a distance of 2.6 AU from the binary pair. This third star (Algol C) orbits the binary in a period of 1.86 years. Now just as Algol A and Algol B orbit around their common centre of gravity (cf. Fig. 4.3), so Algol A + B and Algol C orbit around *their* common centre of gravity. Since the orbit of Algol C is close to the line of sight (but not close enough to cause additional eclipses), the distance from us to Algol A + B changes by 1.3 AU every 1.86 years. The orbital speed of Algol A + B around the centre of gravity with Algol C is thus about 10 km/s. When the motion Algol A + B is more-or less along the line of sight then its distance from Earth alters by about 2.5 million kilometres between one primary eclipse and the next. Light takes about 8 s to cover that distance. Thus whilst the orbital motion of Algol A + B is bringing it towards us the observed interval between primary eclipses will be about 8 s shorter than the orbital period of Algol A and Algol B. Likewise when Algol A + B is moving away from us the observed interval will be 8 s longer than the orbital period. The observed interval between the primary eclipses thus varies from about 2 days 20 h 48 min 49 s to 2 days 20 h 49 min 05 s.

Now imagine Algol C replaced by an exoplanet. The observed intervals between eclipses of Algol A and Algol B will still change, although by a smaller amount than before since the exoplanet will have a smaller mass than Algol C. Nonetheless this thought experiment gives us the entire principle behind detecting exoplanet by timing binary star eclipses – the presence of the exoplanet is betrayed by the small variation in the length of the binary star eclipse period. The time interval required for that small variation to repeat is the exoplanet's orbital period. Since the variation in the length of the binary star's eclipse period arises from the time taken for light to cover a greater or lesser distance to reach us, this approach to exoplanet discovery is sometimes called the light-travel-time (LTT) method. Essentially LTT is just another variant on the Doppler effect.

Jae Woo Lee et al. were the first group to discover an exoplanet from a binary star's eclipse period variations. They announced the discovery in 2009 based upon an 8 year campaign of observations of HW Vir that mainly utilized the 0.6-m telescope at the Soback-san observatory in Korea. Their observations revealed not just one variation in the eclipse period but two – with periods of 15.8 and 9.1 years and amplitudes of 77 and 23 s respectively. The binary system itself comprises two small cool stars in a 2.8-h orbit around each other and some 600 light years away from us. The two objects orbiting the binary turned out to have masses of ≥19 Jupiter masses (HW Vir b) and ≥8.5 Jupiter masses (HW Vir c). Thus only HW Vir c is an exoplanet – HW Vir b is almost certainly a brown dwarf. If the two objects have negligible internal heat sources, then their cloud-top temperatures should be around –40°C and 0°C – but when the brighter of their host stars is eclipsed, these could drop to around –240°C in a few minutes – which should lead to some interesting weather patterns for HW Vir c.

Since HW Vir c, two further exoplanets orbiting binary stars have been discovered, both by Sheng-Bang Qian et al. based upon observations from various telescopes at the Yunnan observatory in China and combined with archive data. DP Leonis b orbits 8.6 AU out from a red-dwarf and white dwarf binary in a period of 24 years and has a mass of ~6.3 Jupiter masses. While QS Vir b, with a similar mass to DP Leo b, orbits its binary host in 7.9 years at a distance of 4.2 AU. The host star in this case, QS Vir, is another red-dwarf and white dwarf binary but it has been suggested that it could also be a dwarf nova that is currently inactive. Should QS

Vir become an active dwarf nova, then the weather on QS Vir b is likely to become even more spectacular than that on HW Vir c.

The detection of exoplanets from the period variations of pulsating stars is no different in principle from that for binary stars except that the pulsation period of the star replaces the interval between successive eclipses as the quantity that changes because of the exoplanet's presence. However, although many stars (Cepheids, Miras, RR Lyrae stars, etc.) vary regularly, most of the variations are insufficiently stable to allow the very small changes resulting from an orbiting exoplanet to be detected. The exception is the class of stars known as the sub-dwarf B pulsators (also called extreme horizontal branch stars). These are hot stars with low masses that may develop after a star has evolved into a red giant and then lost its outer hydrogen layers. Their pulsation periods are in the region of a few hundred seconds or a couple of hours and some have both modes of pulsation at the same time. The stability of their periods is equivalent to gaining or losing a second in 20,000 years.

V391 Peg was found to be a sub-dwarf B pulsator in 2001 with multiple periods around 340–350 s. In 2007 Roberto Silvotti et al. showed that there was a cyclical variation in the star's periods of about ±5 s over 3.2 years. This was due to a ≥3.2 Jupiter mass planet, V391 Peg b, orbiting 1.7 AU out from the star in that same interval. When its host star was a red giant, this planet must have come close to being engulfed by it. Although the radius of the red giant at its maximum would only have been around 0.7 AU, the planet was almost certainly then in a tighter orbit. The effect of mass loss from the star – and V 391 Peg has probably reduced from 0.85 to 0.5 solar masses – is to increase the size of the planet's orbit. So at the time that the host star's radius was 0.7 AU, the exoplanet would have been in a 1 AU orbit. During the day then any unfortunate ETs inhabiting V 391 Vir b would have had the terrifying view of a star filling a third of their sky.

The EXOTIME collaboration (http://www.na.astro.it/~silvotti/exotime/) has been formed recently with the purpose of hunting for exoplanets belonging to sub-dwarf b stars and also to white dwarfs.

Those exoplanets discovered through transits clearly provide a highly stable repetitive phenomenon in the transits themselves.

Careful timing of these may lead to the detection of further exoplanets in a similar fashion to those of binary stars. The new planets may or may not also transit the star. Timing variations in the 1.84-day period of WASP-3 b have recently in this way suggested the presence of a second 15 Earth mass exoplanet in the system although this has yet to be confirmed. Transit timing variations can also be used to confirm the reality of exoplanets that are suspected to exist via other approaches. The two multi-planet systems Kepler-9 (three exoplanets) and Kepler-11 (six exoplanets) were wholly or partially confirmed by this method after their existence had been suspected from the observation of the transits themselves.

Note

1. In August 2011 the discovery of a companion to another pulsar (PSR J1719-1438) was announced. The companion has a mass just larger than that of Jupiter, but like the companion of XTE J0929-314 it is the remnant of what was once a star. The companion is probably a stripped down carbon and oxygen white dwarf. At the pressures found even near the white dwarf's surface most of the carbon is likely to be in the form of diamond - even Elizabeth Taylor never sported a jewel this big!

10. On the Track of Alien Planets – Other Approaches (0% of All Exoplanet Primary Discoveries)

There are a number of searches based upon other ways of detecting exoplanets that have yet to make their first discovery though some have detected already-known exoplanets.

Astrometry

Three or four decades ago most astronomers expected that when the first exoplanet was found, astrometry would be the observational approach making the discovery. Astrometry is the science of measuring the positions and changes in the positions of objects in the sky. It was the method that was used to make the first detection of a star's unseen companion (Sirius B – see below).

In the 1830s astrometry was state-of-the-art science and precision measurements were being made at numerous observatories in the attempt to detect the parallax of stars arising from the Earth's orbital motion and so make the first direct measurement of the distance to a star. Freidrich Bessel working at the Königsberg Observatory (Königsberg was then in East Prussia, it is now known as Kaliningrad and forms part of Russia) was first across the winning line when in 1838 he announced that he had measured the parallax of the star 61 Cygni. His value of 0.314 s of arc placed the star at a distance of 9.8 light years (the modern value is 0.287 s of arc giving the distance as 11.4 light years).

As a part his parallax work Bessel also measured the positions of Sirius and Procyon (α CMa and α CMi) amongst numerous

C. Kitchin, *Exoplanets: Finding, Exploring, and Understanding*
Alien Worlds, Astronomers' Universe, DOI 10.1007/978-1-4614-0644-0_10,
© Springer Science+Business Media, LLC 2012

other stars. For Sirius in particular he noticed that its proper motion of 1.4 s of arc per year was not uniform but oscillated from side to side of the straight-line path by a bit under 2 s of arc over about a 50-year interval (Figure 3.6). In 1844 he predicted that Sirius must have an unseen companion which was orbiting the visible star and whose gravity pulled it from side to side. The companion in this case was not a planet, but another star and it was "unseen" because it was (and is) 8,000 times fainter than the main star (now called Sirius A). In the telescopes of that time the companion (Sirius B) was lost in the glare from Sirius A. However in 1862, whilst testing a newly made 0.47-m telescope objective lens, the Massachusetts optician Alvan Clark saw Sirius B directly. In a similar fashion Bessel suggested that Procyon must have an unseen companion. It took longer for anyone to see Procyon B since it is 14,000 times fainter than its main star, but in 1896 using the 0.9-m Lick refractor, John Schaeberle was successful in observing it.

Both Sirius B and Procyon B are white dwarf stars and so have masses comparable with those of their primary stars (Sirius B has half the mass of Sirius A and Procyon B has a mass of 40% that of Procyon A). The positional changes in the sky of their primary stars, due to their orbits around the common centre of gravity of the system (Figure 4.3), are thus by comparatively large amounts. An exoplanet, even at several times the mass of Jupiter, will have only a few percent of the mass of its host star, nonetheless, as we have seen earlier, there will be an orbital motion of the host star around the common centre of gravity and so a change in the position of the star in the sky. The change in the position of a star in the sky arising from an exoplanet in orbit around it will be much less than the changes for Sirius A and Procyon A, but it is still the detection of this changing position that underlies the astrometric approach to detecting exoplanets.

With a successful record in detecting unseen stellar companions dating back a century and a half therefore, astrometry was widely expected to detect exoplanets. However despite many claims for such detections they have all, to date, been refuted and astrometry has yet to discover its first exoplanet. The claimed discoveries date back to 1855 and continue to be made, with the latest being in 2009. Some of the more notable claims are listed below.

Refuted claims for the discovery of exoplanets by astrometric measurements

Star	Date	Observer	Mass (Jupiter masses)
61 Cyg A	1893	Johannes Wilsing	?
61 Cyg A	1942 and 1957	Kaj Strand	8
61 Cyg A	1977	Alexander Deutsch et al.	6, 12
61 Cyg B	1977	Alexander Deutsch et al.	7
70 Oph	~1790	William Herschel	?
70 Oph	1855	William Stephen Jacob	?
70 Oph	1899	Thomas Jefferson Jackson See	?
70 Oph	1943	Dirk Reuyl and Erik Holberg	10
Barnard's star	1963, 1982	Peter (Piet) van de Kamp	0.5, 0.7
Barnard's star	1973	Oliver Jensen and Tadeusz Ulrych	0.8, 1.4, 1.5, 1.5, 1.6
Lalande 21185	1951, 1960	Peter van de Kamp and Sarah Lippincott	9 to 43
Lalande 21185	1996	George Gatewood	0.9
VB10	2009	Steven Pravdo and Stuart Shaklan	6

Although the latest claim (VB10) has not been confirmed, it is the result of a long-term astrometric search programme with the potential for exoplanet discovery. STEPS (Stellar Planet Survey) uses the Hale 5-m telescope and its long-term positional accuracy is 2 milli arc-seconds.

Astrometry has however had one success that has shown that is a viable method for discovering exoplanets. In 2002 Fritz Benedict et al. using observations made with the HST detected the astrometric motion of the star Gliese 876 caused by its (still unseen) exoplanet, Gliese 876 b. Gliese 876 has a mass a third that of the Sun and is a red dwarf some 15 light years away from us in Aquarius. It is known to host three exoplanets. The outermost one, Gliese 876 b, discovered in 2000 by the radial velocity method, is 0.2 AU out from the star and has a mass of 2.6 Jupiter masses. The orbit of the star around its centre of gravity with Gliese 876 b is thus about 0.003 AU in diameter – corresponding to a movement in the sky of 0.5 milli arc-seconds.

A radio astrometric study of cool dwarf stars (RIPL) is currently being undertaken though this has yet to produce any results. While in the future the spacecraft, Gaia and the Stellar Interferometry Mission (now known as SIM Lite) are expected to achieve positional accuracies of a few millionths of an arc second and so be able to detect Earth-mass planets out to a distance of 30 light years.

Polarimetry

Almost all the light emitted by most stars is unpolarized. Only on a small scale (by stellar standards) do phenomena such as active region magnetic fields and flares lead to some emission of polarized light. Even these result in only very tiny polarizations in the integrated light that we receive from the star. However when the light from the star is scattered in the atmosphere of an exoplanet, the scattered radiation is polarized and strongly so in some directions.

When we observe a star together with an exoplanet that has an atmosphere, except in a very few cases where we can see the planet separately, the light that we receive is a combination of that emitted by the star and that scattered from the planet. When the contribution from the planet is polarized, so also will be the integrated light that we receive – but for a typical Jovian planet the polarization will be only about one part in a million. Nonetheless it is by detecting that polarization that some planet hunters hope to discover exoplanets. In this they should be aided by the polarization changing as the planet orbits its star.

The polarization is measured using an instrument called a polarimeter and this is essentially a device for measuring the brightness of the star when seen through a polarizer. A polarizer is an optical component that transmits light polarized in one direction and is opaque to light polarized in the perpendicular direction. Polaroid™ is a well known polarizer much used in sunglasses, but most astronomical polarimeters employ other devices, such as Wollaston prisms, for the purpose. The polarizer is rotated within the polarimeter and if the incoming light is polarized the output from the instrument will vary cyclically at twice the speed of rotation. Although simple in concept, polarimeters have to be very sophisticated instruments in practice if they are to detect levels of polarization of one part in a million.

HD 189733 A is a smallish cool star in Vulpecula about 64 light years away from us. It forms a binary star system with HD 189733 B some 200 AU away from it and the two stars complete an orbit every 3,200 years. In 2005 HD 189733 was discovered to have a transiting exoplanet in a 2.2 day orbit with a separation of just 0.03 AU. HD 189733 b is a large hot Jupiter and is the only exoplanet to have been mapped so far (Figure 7.5). In 2007 a team of astronomers from Switzerland and Finland led by Svetlana Berdyugina used a polarimeter on the Finish 0.6-m KVA telescope (Kungliga Vetenskapsakademien) on La Palma to observe the star in the blue part of the spectrum. They detected the polarization component from the planet and found that it peaked at levels of 200 parts per million when the planet was at its maximum separations from the star (as seen from Earth). These positions correspond to the planet having a "half-moon" appearance were we able to see it in sufficient detail. From the polarimetric observations HD 189733 b was found to have a radius of 1.5 Jupiter radii – 30% larger than that suggested by the transit observations. The difference in the radius detected by the two methods is almost certainly due an extensive outer atmosphere that scatters the light from the star so producing the observed polarization but which does not absorb significantly when it is seen in silhouette during a transit.

So far HD 189733 is the only star where exoplanet-induced polarization has been detected. Several groups though are working on instruments with the potential to measure polarizations at the level of a few parts per million. The Swiss team involved in the HD 189733 observations and based at the Zurich Eidgenössische Technische Hochschule has developed ZIMPOL (Zurich Imaging Polarimeter) for use on ESO's VLT. This instrument has reached a level of precision in measuring polarization of about ten parts per million. PlanetPol (not an acronym) developed by a team from the University of Hertfordshire and led by Jim Hough improves on ZIMPOL's performance by a factor of ten. Unusually given the almost universal use of CCDs it employs avalanche photodiodes as its detectors. It has been used on several telescopes including the 4.2-m William Herschel telescope. Observations of the exoplanet host stars 55 Cnc and τ Boö were made between 2004 and 2006. 55 Cnc has five exoplanets with the largest, 55 Cnc d, being a cold Jupiter (orbital radius 5.8 AU, mass ≥3.8 Jupiter masses). Closest to the star at 0.038 AU is 55 Cnc e with a mass around 0.024 Jupiter masses (8 Earth masses). τ Boö has one exoplanet, a hot Jupiter in a 0.046 AU orbit and with a mass of ≥3.9 Jupiter masses. Despite the high level of precision in the polarization measurements, PlanetPol's results did not show any evidence of exoplanetary polarization for either star.

Circumstellar Disks

Perhaps up to 15–20% of all stars are embedded in large tenuous clouds of gas and dust particles. The clouds are usually rotating and so take up the shape of lentils or of an athlete's discus and are commonly referred to as circumstellar disks. The star is usually more-or-less at the centre of the disk though this is not always the case. The material in the disks is thought to be closely associated with the formation of planets although whether it provides the building blocks for the planets or is the debris from planetary collisions is not always clear. Further discussion of this aspect of circumstellar disks and of their properties is left until Chap. 13.

Circumstellar disks are included in this chapter because they can indicate the presence of one or more exoplanets orbiting their central star. The gravitational field of an exoplanet within a circumstellar disk is likely to clear the disk's material from a region around the planet's orbit. The planet does this either by capturing the material itself, and so growing in mass, or by perturbing the orbits of the particles so that they move away from the region, or most probably via both mechanisms. The presence of a low or high density ring or rings within a disk may thus indicate that an exoplanet is present. Often the structure takes the form of a central clearing so that the disk has the shape of a ring doughnut. Central clearings are to be found in around a third of circumstellar disks.

Structures within circumstellar disks indicate the possible presence of an exoplanet but the planet must be found by other methods. As an example of this in 1983 the IRAS spacecraft detected an infrared excess coming from the star β Pic that was attributed to cool matter surrounding that star. In the following year a direct visible light image obtained by Bradford Smith and Richard Terrile using the du Pont 2.5-m telescope confirmed the presence of a circumstellar disk some 3,000 AU across (Figure 3.12). Many subsequent observations have shown structures within β Pic's disk. The disk itself is asymmetric and there are elliptical rings 500–800 AU out from the star. The central region of the disk is inclined at an angle of about 5° to the outer part and concentrations of material are found at distances of 6, 14, 28, 52 and 82 AU from the star. Speculation that at least some of these structures could be due to one or more exoplanets (the larger scale structures seem likely to be due to a passing star) had been going on for some time when, in 2008, a direct image of a possible planet was obtained using the VLT by Anne-Marie Lagrange et al. two years later it was confirmed to be an exoplanet when, having moved somewhat around its orbit, it became visible on the other side of the star (Figure 7.3d). The planet is in a ~17 to ~35-year orbit 8–15 AU out from its host star with a probable mass of 8 Jupiter masses.

ε Eri is another star with structure in its circumstellar disk. It has a central cleared region, clumps of denser material and the disk is asymmetric. In 2000, from radial velocity measurements, the star was announced to possess a 1.5 Jupiter mass exoplanet in a roughly 7 year orbit, 3.5 AU out from the star, but there is still some uncertainty about this. The disk structure suggests a possible second planet with an orbital radius around 40 AU and an orbital period in the region of three centuries, but so far this planet has not been directly detected. Infrared observations by the Spitzer spacecraft suggest the presence of two asteroid belts. If this is correct their existence would make the likelihood of larger planets seem to be more probable.

Alien astronomers observing the present solar system from a distance would probably be able to infer the presence of the planet, Neptune, from the gap that it creates in the inter-planetary dust cloud originating from debris due to collisions between KBOs.

Spitzer observations of a circumstellar disk for the star HD 172555 also provide evidence for exoplanets but in a rather different fashion from ε Eri. The infrared spectrum obtained by the spacecraft suggests that the disk contains melted glass and silicon monoxide gas. This in turn suggests the presence of vapourized rock in the disk and that the disk is formed from the debris of a collision between two exoplanets. If correct the two planets could have been about the size of Mercury and the Earth's moon and the collision might have occurred about 1,000 years ago.

White Dwarf Atmospheres

The Spitzer telescope has been involved in another investigation suggesting the one-time existence of now destroyed exoplanets. The work though probably implies that there are also other, still existing, exoplanets to be found. White dwarfs are the end points of the evolution of solar-type stars. They have thin atmospheres that are thought to be almost pure hydrogen or pure helium, but the spectra of white dwarfs sometimes show the presence of other elements such as calcium and magnesium. Jay Farihi et al. using data from the Sloan Digital Sky Survey have recently suggested that the proportion of white dwarfs with such contaminated atmospheres could be as high as 20% and that the inter-stellar medium was unlikely to be the source of the contaminants. The team also used Spitzer in 2009 to determine that between 1% and 3% of white dwarfs are embedded in small dusty regions very near to the stars. They suggest that the white dwarfs with contaminated atmospheres have received the contaminating material from their surrounding dusty regions. They go on to suggest that the dust originates from asteroids that have approached too closely to the white dwarf and been broken up by tidal forces.

In 2010 a cool, faint white dwarf in Gemini was found with the highest levels of contaminants to date. If due to an asteroid break-up and assimilation by the white dwarf, then the contaminating object must have had a mass at least as large as the solar-system dwarf planet, Ceres (diameter 940 km, mass 0.0002 Earth masses), and a composition similar to that of the Earth. Clearly these exo-asteroids are no longer in existence. However something must have caused them to approach the white dwarf sufficiently closely to be broken up and one possible "something" could be gravitational perturbations by larger exoplanets further out from the star.

Like the structure in circumstellar disks, the contamination of white dwarf atmospheres by elements such as calcium and magnesium is only an indicator of where exoplanet hunters using more direct methods of detection might search. Perhaps more importantly though some 90% of stars end up as white dwarfs, if somewhere between 1% and 20% of white dwarfs possess exo-asteroids and perhaps exoplanets then it implies that a similar or larger proportion of most stars must also possess exo-asteroids and exoplanets.

Radio-Based Methods

The majority of studies of exoplanets so far have been in the visible and infrared parts of the spectrum. As we have seen though, the very first exoplanets to be found were detected through radio observations of the emissions from a pulsar. It is possible that exoplanets could be detectable at radio frequencies by other methods though this has yet to happen.

Exoplanets could be directly detectable from their own radio emissions. Jupiter emits strongly in the low frequency radio region due to electric currents flowing in the planet's magnetosphere. We may expect that at least some of the jovian-sized exoplanets will have similar or stronger emissions. Present-day radio telescopes are potentially just able to detect Jupiter-strength emissions out to a distance of a few tens of light years. Foreseeable improvements in receivers and new radio telescopes under development should make the direct radio detection of exoplanets a more viable prospect within a few years. Another possibility is the detection of artificial radio emissions from the ET inhabitants of an exoplanet. However SETI (Chap. 14) has conducted intensive searches for such emissions for many years now without success.

Alternatively exoplanets could be detectable by observing the radio emissions from their host stars and using techniques such radial velocity variations, transits or changes in the stars' positions in the same manner as already discussed for optical observations. Currently the Radio Interferometric Planet Search (RIPL) using the 100-m Green Bank radio telescope and the Very Large Array (VLA) is conducting a 3-year study of cool dwarf stars in an attempt to detect exoplanets from changes in the stars' positions. Radio positions of the stars can be measured to a precision of a tenth of a micro-arc-second while a Jupiter sized exoplanet in a 1 AU orbit around a small star would produce a displacement in the star's position of a micro arc-second at a distance of 15 light years. Currently though, RIPL has only set upper limits to the possible motions of its chosen stars.

11. Where Do We Go from Here? – Future Approaches to Exoplanet Detection and Study

The investigation of exoplanets has so far concentrated simply on just finding them. Now, however, with over 500 known exoplanets, the emphasis within the subject is changing towards studying the details of the individual planets and trying to understand their properties as a group. Of course, efforts to discover new exoplanets will continue alongside these follow-up studies of already known exoplanets and one important aim must be to reduce the present in-built bias towards finding hot Jupiters.

The outlook for exoplanet detection over the next few years to a couple of decades from now divides into two possibilities:

1. Innovative and different approaches based upon new physical principles and
2. More of the same, but better/faster/more precise/more detailed/more stars and planets/etc.

Innovative and Different Approaches Based Upon New Physical Principles

The prospects for *completely* new physical principles are clearly unknowable and there are no obvious new ways to discover exoplanets in addition to those already in use. Some radical improvements to those existing methods can be foreseen such as the use of laser combs (see below) as comparison spectra for radial velocity measurements and the increased use of interferometry for direct

C. Kitchin, *Exoplanets: Finding, Exploring, and Understanding Alien Worlds*, Astronomers' Universe, DOI 10.1007/978-1-4614-0644-0_11, © Springer Science+Business Media, LLC 2012

imaging. Where we may expect to see new (to exoplanet study) methods being applied is in the investigation of the natures of the exoplanets. In particular, spectroscopy at wavelengths from the near ultra-violet to the radio region is likely to start to reveal the compositions and structures of exoplanets' atmospheres and surfaces, their temperatures, rotational velocities, etc. We may also expect more exoplanets to be mapped at low resolution by indirect methods such as that used for HD 189733 b (Figure 7.5).

In 2005 Theodor Hänsch and John Hall part shared the Nobel Prize for physics for their development of lasers able to emit their light in the form of pulses with extremely short durations. One application of such lasers is the production of a spectrum that comprises a series of very sharp emission lines with a constant separation in frequency. Such a spectrum is called a laser comb (Figure 11.1) and it can be used as the comparison spectrum (Figure 5.1) for determining radial velocities. Since the laser comb can be made extremely stable for long periods of time and because of the large number of lines present in it, the radial velocity of a star can potentially be measured to an accuracy of ±10 mm/s or better (the current state-of-the-art is about ±1 m/s). A laser comb comparison spectrum is currently being developed for the HARPS spectrograph. A second approach to improving the accuracy of radial velocity measurements has already been mentioned (see Exoplanet Tracker). This is the combination of an interferometer and a spectrograph and both this approach and the laser comb are likely to be used much more in the near future.

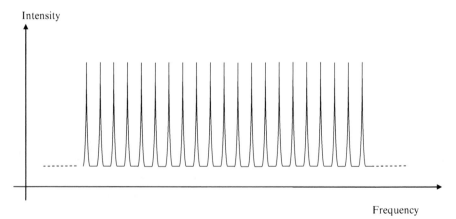

FIGURE 11.1 Intensity variations in a laser comb spectrum.

A possible indicator of the existence of exoplanets around white dwarfs has already been discussed and this is the presence of contaminants such as calcium and magnesium in their atmospheres. A similar indicator, but for solar-type stars, could be the low abundance of lithium. A statistical study of 500 such stars, of which 70 were known to possess exoplanets, has shown that the stars with exoplanets had much less lithium in their surface layers than those without exoplanets. The reason for the absence of lithium is not yet known, but the fact of the correlation of its absence with exoplanets can still be used to suggest good candidate stars for exoplanet hunters to examine. Also for solar-type stars (spectral classes F, G and K – see Appendix IV for a brief summary of stellar spectral and luminosity classification) there is an observed correlation between the presence of exoplanets and the abundance of the heavier elements, particularly iron, in the host star. About 25% of stars with high iron abundances have giant exoplanets compared with around 3% of those at the low abundance end of the scale. An obvious, though not necessarily correct, explanation for this latter correlation would arise if, as seem likely, stars rich in the heavier elements formed from inter-stellar gas clouds that were also rich in the heavier elements. There would thus be plenty of heavy-element compounds in the proto-stellar disk to condense out and form the rocky/metallic cores that probably act as nuclei for giant planet formation.

An indicator that may help to detect exoplanets with liquid oceans – whether these be composed of water, ammonia, methane or some other compound – is that we may expect reflections off liquid surfaces that will be different from those from solid or gaseous layers. In particular, liquids will reflect brightly (glint) at angles where a mirror would reflect the light. Such glints could double the brightness of the exoplanet and produce a characteristic variation of the planet's brightness with its phase change as it orbits its host star. Along the same lines it is possible that we might be able to detect the presence of the alien equivalent of forests. The way in which the brightness of a forested exoplanet varies as it orbits its star would differ from the variations for one with low-growing or no vegetation because of the shadows cast by the alien trees.

The technique of interferometry has been briefly mentioned earlier. It is a method of dramatically increasing the resolution and sometimes the sensitivity of telescopes inexpensively. It was first used extensively with radio telescopes, but is now used at all except the shortest wavelengths.

An interferometer splits light into two beams and then recombines them, but with one of the beams having travelled further than the other. At some points in the recombined image the light in the two beams will have been shifted by a whole number of wavelengths so that the wave crests in one beam align with the wave crests in the other beam (and wave troughs align with wave troughs). The two light beams thus reinforce each other and at that area of the image there is a bright spot. In other areas of the image the crests of the light waves in one beam will align with the troughs in the other beam and they will cancel each other out – resulting in a dark spot in the image. The bright and dark areas, or fringes as they are normally called, form a regular alternating pattern.

A basic astronomical interferometer consists of two telescopes whose outputs are combined. The resulting image consists of a pattern of bright and dark fringes. The fringe pattern looks nothing like a "proper" image of a star or planet (or whatever) – but the details of the fringes can be used to reconstitute the proper image at a much higher resolution than that provided by either of the telescopes. The resolution in fact depends upon the separation of the telescopes – so two telescopes 100 m apart are the equivalent to a single 100-m telescope. By using many telescopes and by observing over a period of 12 h it is possible to build up the sensitivity of a 100-m telescope as well as its resolution. Many optical telescopes such as the VLT and Keck instruments now frequently operate as interferometers. As we have seen earlier, the direct observation of exoplanets requires the use of coronagraphs (Appendix IV) to blank out the overwhelming light from the host star. With an interferometer it is possible to arrange that the star occupies the position of one of the dark fringes while the exoplanet is within one of the bright fringes - so performing the same function as a coronagraph. An interferometer used in this fashion is called a nulling interferometer (because the star's image is deleted or nulled).

Fresnel lenses are widely used in car headlights and over-head projectors because although the image quality is poor they are cheap to make. They are best imagined as though a conventional converging lens with one flat and one curved surface has had most of its interior hollowed out and then the remaining curved surface cut into narrow rings and mounted onto a flat plate. Since refraction of light takes place only at the surface of the lens, the loss of the interior has little effect upon the imaging properties of the device. A similar lens can be made out of opaque segments and which operates through diffraction rather than by refraction. A Fresnel lens operating by diffraction could be made very large and very light and could therefore be launched into space where it could produce direct images of exoplanets.

More of the Same, but Better/Faster/ More Precise/More Detailed/More Stars and Planets/etc.

Significant improvements in the detectors used on telescopes to detect and image exoplanets are unlikely since these devices are already of a high standard – CCDs for example pick-up 90% of the light that falls onto them. Most detectors now take the form of arrays of individual point-source detectors and the largest such arrays are currently $2,048 \times 4,096$ in size – a total of over eight million individual detectors. While it is possible that manufacturing techniques may produce larger individual arrays, the present approach of stacking many arrays side-by-side in order to image a larger field of view seems likely to be the way things will progress in the near future. The Kepler spacecraft's detector mosaic for example comprises 42 $1,024 \times 2,048$ pixel CCDs (Figure 6.10) while the Sloan digital sky survey telescope used 30 $2,048 \times 2,048$ arrays. Making larger mosaics of arrays is "just" a matter of finding more money.

Improvements to detectors may come about through the reduction in their background noise levels. In particular detectors that count individual pulses (photons) of light will reduce the noise involved in picking up the signal from the detector – known as read-out noise – to zero. The manufacture into arrays of one type zero-read-noise detector, known as an avalanche photo-diode, is currently in its early stages.

Improvements to existing methods of discovering and studying exoplanets range from instruments currently under construction to a plethora of "concept studies" that comprise little more than a good acronym and a nice "artist's impression." Some developments seem more promising, but even promising developments can fall victim to changes of policy, changes of government, currency fluctuations or disasters in the financial markets, so some, perhaps many, of these proposals may yet fail. Amongst the more promising developments we have the following possibilities.

Large telescopes using sophisticated adaptive optics systems are now able to deliver positional measurements to an accuracy better than 0.2 milli arc-seconds. As longer term data bases of such measurements are built up over the next few years, it is likely that the detection of exoplanets by astrometry will finally be successful.

The unique flying observatory, SOFIA (Stratospheric Observatory For Infrared Astronomy – Figure 11.2) has recently started operations. It uses a 2.7-m telescope that observes through a hole cut in the side of Boeing 747 jet aircraft which flies at a height of 14 km in order to be above 99% of the water vapour (an infrared absorber) in the Earth's atmosphere. While it is possible that SOFIA may be

FIGURE 11.2 SOFIA (Reproduced by kind permission of NASA and Jim Ross).

able to observe exoplanets directly, it is more likely that it will be used to study the details, including the mineralogy, of the disks of gas and dust around stars within which exoplanets may be forming or have recently formed or which are the debris from collisions between exoplanets. It will also provide data on the planets within our own solar system for comparison with similar exoplanets.

A proposed replacement for HARPS on the VLT is ESPRESSO (Echelle Spectrograph for Rock Exoplanet and Stable Spectroscopic Observation). A preliminary study of the instrument was completed in 2010. It is to operate in the visible region and with the aim of being able to measure radial velocities to a few tens of milli-metres per second. It will be able to use light from all four of the 8-m telescopes of the VLT – equivalent to a single 16-m telescope. The Spectro-Polarimetric High contrast Exoplanet Research instrument (SPHERE) also for the VLT is expected to start operating in 2012. SPHERE will include adaptive optics to correct for atmospheric distortions to a high level, coronagraphs, an SDI (Appendix IV), a spectrograph and a polarimeter. It is expected to be able to image directly tens, perhaps hundreds, of giant exoplanets. The Gemini Planet Imager (GPI) is being built for the 8.1-m Gemini South telescope based upon the existing Near Infrared Coronagraphic Imager (NICI) and it too is expected to be completed in 2012. It is a high quality adaptive optics system combined with coronagraphs that should not only be able to image exoplanets directly but also obtain their spectra.

There are several proposals for extremely large ground-based telescopes. Although some are still in the planning stages, it seems likely that at least one will eventually be constructed. The Carnegie Institution's Giant Magellan telescope (GMT) will use seven 8.4-m mirrors on a single mounting with the equivalent area of a single 22-m mirror. Work has already started on the instrument with the casting of some of the mirrors and it is about 40% funded at the time of writing. Completion is expected around 2018. The U.S.A. and Canada's Thirty Meter Telescope (TMT) is currently in the planning stages and has had around $250 million committed to it so far. Completion is possible around 2025. The ESO's European Extremely Large Telescope (E-ELT – Figure 11.3) is to be a 42-m diameter instrument, also currently in the planning stages, although a completion date of 2018 is expected if funding can be

FIGURE 11.3   An artist's concept of the European extremely large telescope.
(Reproduced by kind permission of ESO).

found. Once built such telescopes are likely to contribute to most methods of discovering and studying exoplanets. Using highly sophisticated adaptive optics they will have resolutions better than 10 milli-arc-seconds. Many exoplanets will therefore be able to be imaged directly. With spectroscopes, radial velocities should be measureable to ±10 mm/s – enabling Earth-twins in habitable zones to be found. Spectroscopy of the atmospheres of exoplanets both directly and during transits when the atmosphere is back-lit by the star will provide details of the elements and molecules making up those atmospheres. Polarimetry should similarly be able to characterize the surfaces of rocky and icy planets. Fine details of proto-planetary disks should also be observable.

The majority of anticipated developments in exoplanet hunting however are likely to be space-based. NASA's James Webb space telescope is a partial replacement for the HST and is due for launch in 2014. It will have a 6.5-m diameter main mirror made up from 18 hexagonal segments and will operate primarily in the infrared. For exoplanets the JWST is likely to concentrate on follow-up studies.

Its various instruments will enable it to undertake direct imaging using coronagraphs, to make precision observations of transits and via spectroscopy to study exoplanetary atmospheric and surface compositions and structures. SPICA (Space Infrared telescope for Cosmology and Astrophysics) is a planned ESA mission that is currently undergoing industrial assessment studies. It is to operate in the mid to far infrared region using a 3.5-m mirror and carry spectrographs and coronagraphs. If it is given the go-ahead launch could be in 2017 or 2018.

NASA's SIM Lite (a reduced-cost version of the original SIM – Space Interferometry Mission) is currently awaiting further funding and is planned to use two 0.5-m telescopes separated by 6 m to measure the positions of stars to within 1 micro arc-second (the angular size of this capital "O" were it to be on the Moon and viewed from the Earth). If it goes ahead, SIM Lite could detect nearby Earth-mass exoplanets from the wobbles in their host stars' positions. Gaia is an ESA mission planned for launch in 2012 that will measure the positions of a 1,000 million stars and galaxies to an accuracy of 6 micro arc-seconds. It is estimated that Gaia could potentially detect hundreds of thousands of Jupiter mass exoplanets during its 6 year lifetime.

The international Atacama Large Millimeter Array (ALMA) is now under construction high in the Chilean Andes. It is to consist of 66 or more 7 and 12-m parabolic dishes observing as an interferometer in the 0.3–3.6 mm wavelength region. When fully completed in 2012 it will be able to detect new and still-forming giant exoplanets directly and all types of giant exoplanets indirectly through astrometric measurements. It will also be able to observe fine details of proto-planetary disks around stars. At longer wavelengths the Square Kilometer Array (SKA) is in the early stages of planning with a completion in 2024. It is envisaged as a collection of small radio aerials spread over an area up to 3,000 km across and with a total collecting area of a square kilometre. It should be able to detect the radio emissions from giant exoplanets directly.

Other schemes for studying exoplanets including concept studies and proposals currently in the early stages of planning are listed below.

Instrument	Ground or space-based	Detection method	Summary	Status
ACCESS – actively corrected coronagraph concepts for exoplanetary system studies	Space	Direct imaging, spectroscopy	1.5-m telescope operating in the visible region	Concept 2009
Antarctic plateau interferometer	Ground	Direct imaging	Simple interferometry	Concept 2005
ATLAST – advanced technology large-aperture space telescope	Space	Direct imaging, transits, spectroscopy	8 to 16-m telescope with a coronagraph	Concept 2009
BOSS – big occulting steerable satellite	Space	Direct imaging	A large telescope using a coronagraph with an external stop positioned 100,000 km away	Concept 2000
DARWIN	Space	Direct imaging	An ESA study for a flotilla of four or more spacecraft carrying 3 to 4-m telescopes acting as interferometers	Study completed in 2007
EUCLID	Space	Gravitational microlensing	A 1.2-m telescope operating in the visible and near infrared	Continuing development studies by ESA
FKSI – Fourier-Kelvin stellar interferometer	Space	Direct imaging	Several spacecraft carrying small telescopes that act as a nulling interferometer	Concept 2009
Fresnel imager space mission	Space	Direct imaging	A large (up to 100 m in diameter) Fresnel diffraction lens feeding images to a separate spacecraft	Concept 2005
GEST – galactic exoplanet survey telescope	Space	Gravitational microlensing, transits	A NASA study for a spacecraft-borne 1.5-m telescope observing the galactic bulge	Concept 2000

Name	Location	Method	Description	Status
Hyper telescope	Space	Direct imaging	Up to 150 3-m telescopes separated by up to 100 km, kept in position by lasers and using nulling interferometry to resolve Earth-sized exoplanets	Concept 2009
HZPF – habitable zone planet finder	Ground	Radial velocity – near infrared	A fibre-optic fed near infrared spectrograph for the Hobby-Eberly telescope capable of measuring radial velocity to an accuracy of ±1 to ±3 m/s. Aims to detect low mass exoplanets around M-type dwarf stars	Proposed 2009
New worlds discoverer/observer/imager	Space	Direct imaging	A variety of related proposals for imaging exoplanets using an external stop coronagraph. The suggestions range from simply putting a stop for an existing or planned telescope (e.g. the JWST) to use through to a dedicated pair of spacecraft to a fleet of spacecraft and stops	Concept 2005 – still under study by NASA
PECO – pupil mapping exoplanet coronagraphic observer	Space	Direct imaging	1.4-m telescope with a coronagraph operating in the visible region	Concept 2009
PEGASE	Space	Direct imaging	A CNES study for three spacecraft carrying 0.4-m telescopes acting as simple and nulling interferometers and operating in the near infrared	Concept 2005
PLATO – planetary transits and oscillations of stars	Space	Transits	An ESA candidate mission. 42 small telescopes to cover a 28° × 28° degree field aimed to determine the characteristics of exoplanets especially Earth-twins in habitable zones	Assessment study currently underway

(continued)

(continued)

Instrument	Ground or space-based	Detection method	Summary	Status
SEE – super earth explorer	Space	Spectroscopy, polarimetry	Study of chemical composition of exoplanet atmospheres and surface properties using a >1.5-m telescope.	Concept 2009
THESIS – terrestrial and habitable zone exoplanet spectroscopy infrared spacecraft	Space	Spectroscopy	Molecular spectroscopy from the visible to the far infrared using a cooled 1.4-m telescope	Concept 2009
TPF – terrestrial planet finder	Space	Direct imaging	Possibly several spacecraft forming a nulling interferometer or a large optical coronagraph	Postponed indefinitely
UMBRAS – umbral missions blocking radiating astronomical sources	Space	Direct imaging	Telescope with an external stop coronagraph	Concept 2000

Finally – the *real* question – when are we going to find a twin for the Earth? It is not sufficient just to find an exoplanet with a mass close to that of the Earth – that seems likely to happen before 2015 and something as inhabitable as, say, Mars, before 2020. It is now becoming clear, though, that the conditions for a true Earth Twin (i.e. an exoplanet that we could go and live on immediately without any significant problems) are very stringent and complex. Apart from a suitable mass, the exoplanet also has to be within the habitable zone of the space around its host star and have a nearly-circular orbit (the Earth could be in an elliptical orbit that went nearly as close to the Sun as Venus and as far out as Mars and still be within the Sun's habitable zone – but we would find the temperature variations impossible to live with). It will also need a suitable composition including liquid water and an oxygen atmosphere. The latter would almost certainly mean that photosynthesizing plant life must have evolved on the exoplanet and therefore that it has been in its favourable orbit for thousand million years or more. Possibly also geological processes such as plate tectonics are required to create suitable surface conditions (mountains, valleys, plains, volcanoes, rivers, seas, oceans, etc.). Given these and doubtless other requirements of which we are unaware at the moment, we will perhaps need to find at least a hundred Earth-mass exoplanets before one crops up that is completely suitable. Although Earth-mass planets are probably at least as common as giant planets, the latter are much easier to detect. Perhaps therefore 10,000 exoplanets will have to be discovered before we know of 100 with masses close to that of the Earth. So (Figure 3.18) it is probably going to be 2020 to 2045 before an Earth twin is finally found.

12. Exoplanets Revealed – What They Are *Really* Like

Introduction

The characteristics, structures and constitutions of just the eight objects within the solar system now recognized as planets varies enormously – small/large – solid/gaseous – hot/cold – atmosphere/no atmosphere – rocky/icy – etc. When we add dwarf planets, asteroids, natural satellites, comets, Kuiper belt objects and the rest, the variety becomes truly amazing. Yet the solar system is just 1 of 400 now discovered planetary systems – most of which we already know to be quite different from our own neighbourhood.

Yet even within the solar system there are large blank areas of ignorance, and what are worse, mistaken ideas. What do we understand, for example, about what is going on 10,000 or 50,000 km below the visible cloud tops of Jupiter, or at Pluto's centre, or during massive asteroid impacts such as those that produced the Moon's Mare Imbrium and Mercury's Caloris Basin, or why Europa's surface is a network of cracks and so-on? In fact we are still pretty uncertain about what there is inside the Earth just a few hundred kilometres below our feet.

In this context it is worth remembering some of the mistakes made in the past: Martian canals, Mercury's rotational period equaling its orbital period, the 'discovery' of a planet, Vulcan, inside the orbit of Mercury, the 'observation' of Neith, a non-existent natural satellite of Venus and Themis, a similarly fictional satellite of Saturn and so on. These mistakes arose when solar system objects were studied at or beyond the then state-of-the-art. If such questions and errors remain for intensively observed and nearby objects, what hope have we of saying anything pertinent about exotic planetary systems hundreds or thousands of light years away, just the detection of which strains our current capabilities to the limit?

C. Kitchin, *Exoplanets: Finding, Exploring, and Understanding Alien Worlds*, Astronomers' Universe, DOI 10.1007/978-1-4614-0644-0_12, © Springer Science+Business Media, LLC 2012

Well – we do have some things going for us. Despite our ignorance about some aspects of solar system objects, there are other aspects about which we know a great deal. Thus we can draw analogies and extrapolate from that data. Secondly the laws of physics, chemistry, geology, meteorology, etc. as we have deduced them here on Earth are probably true throughout the universe, at least until you get back to within a tiny, tiny fragment of a second of the big bang, Certainly such laws apply throughout the 'small' volume of the universe occupied by the Milky Way galaxy. Thirdly the detection of an exoplanet usually provides us with an estimate of its mass and distance from its host star, the properties of the host star and sometimes of planet's radius – and you can infer a lot from that. Fourthly the sheer number of exoplanets now known allows us to make inferences about which are common processes and characteristics and which are rare ones – something for which the solar system provides too small a statistical sample. Finally, of course, our observing techniques are improving all the time and one day we *will* be able to see directly the continents, clouds, seas and oceans of a twin Earth.

In fact we have already implicitly made some deductions about the natures of exoplanets, when earlier we divided them up into Hot Jupiters, Hot Neptunes, Cold Jupiters, Super Jupiters, Super Earths, exo-Earths and Free-floating planets. A classification of gas giant planets produced on theoretical grounds before any exoplanets had been discovered is used occasionally. There are five Sudarsky classes whose properties are summarised below. More recently two classes based upon the properties of their atmospheres have been suggested for hot Jupiters. Introduced in 2008 by Jonathan Fortney et al., they are labelled pM and pL. The pM class of hot Jupiters are hotter than the pL and their atmospheres contain the molecules titanium oxide and vanadium oxide (TiO and VO) in gaseous form which strongly absorb the light from their host star. In the cooler pL class of hot Jupiters, the titanium and vanadium have condensed out to become solids and sodium and potassium are the predominant absorbers. The 'M' and 'L' in this notation were chosen by analogy with the M and L spectral classes of cool stars.

Sudarsky classification of gas giant planets and exoplanets. (NB the *main* gases making up the atmosphere in each case are hydrogen, helium and methane – classes I, II and III, or hydrogen, helium and carbon monoxide – classes IV and V).

Sudarsky class	Surface temperature (°C)	Defining cloud component(s)	Example(s)
I	<–120	Ammonia	Jupiter, Saturn, 47 UMa c
II	~–20	Water vapour	HD 28185 b, γ Cep A b
III	80–530	Cloud-free (featureless blue appearance arising from scattering and absorption by methane)	Gliese 876 b, υ And c
IV	~900	Sodium and potassium	55 Cnc b, HD 209458 b?
V	>1,100	Silicate compounds and iron vapour	51 Peg b?

Hot Jupiters

Since hot Jupiters (or Sudarsky Class IV and V planets) were the first exoplanets to be found orbiting normal stars, we will begin by considering what they may be like.

Let us set the scene by starting with Jupiter itself. Jupiter's visible surface is not solid, but is the top of an extremely deep atmosphere largely composed of hydrogen and helium. The temperature at the cloud surface is around –140°C. Meteorological processes akin to the jet streams in the Earth's atmosphere cause it to have a banded appearance (Figure 12.1) and there is also the semi-permanent Earth-sized feature called the great red spot. The planet has a mass two and a half times that of all the other planets added together (0.001 solar masses, 320 Earth masses), it is 143,000 km in diameter and rotates in just 9 h 50 min producing its quite noticeable squashed-sphere appearance. At the last count it had 63 natural satellites.

Our telescopes can see only the surface of Jupiter, so we have to rely on a few indirect clues along with computer modelling to try to guess the planet's structure. The main things we have to go on are its mass and radius (which give its mean density as 1.3 times that of water), its magnetic field (which is 14 times as strong

FIGURE 12.1 A monochromatic image of Jupiter obtained using ESO's VLT in 1998. Jupiter's natural satellite, Io, is just starting to transit the planet's disk (*white spot, bottom left*) while the *Great Red Spot* is on the right hand limb in the main southern atmospheric band. (Reproduced by kind permission of ESO).

as that of Earth), the structure of its gravitational field and the slightly surprising observation that it is radiating away twice as much energy as it receives from the Sun.

Obviously a lot of heat is percolating outwards from Jupiter's central regions. Our best estimate for its internal structure says that it continues to get hotter as you move down towards the centre of the planet. At a depth of about 15,000 km it is around 10,000°C. At 60,000 km from the surface it approaches 20,000°C, and this rises to perhaps 25,000–35,000°C at the very centre.

Most of Jupiter is hydrogen. About a tenth of the planet is helium and all the remaining 90 elements add up to just a few percent. In the outer layers of the planet, from the cloud tops down to about 15,000 km, the hydrogen and helium behave like normal gases. From there to about a depth of 60,000 km, the overwhelming pressure forces the hydrogen to behave a bit like a liquid metal. In particular the hydrogen becomes able to conduct electricity (and it is electric currents flowing in this region that power Jupiter's strong magnetic field). Right in the very centre of the planet, the material may become richer in the elements beyond helium since the weight of these elements and their compounds will tend to cause them to sink downwards. There may thus be a core comprising the elements such as carbon, nitrogen, oxygen, silicon, calcium and iron which form the bulk of the Earth. Probably though there will still be a lot of hydrogen and helium incorporated into the mix. This core could be up to ten times the Earth's mass (3% of the mass of Jupiter) – or if the inside of the planet has been highly turbulent, it may not exist at all.

As well as Jupiter's central temperature being five or six times higher than that at the surface of the Sun, the pressure there is some 40 million times that of our own atmosphere at the sea level. Probably the centre of Jupiter contains a lot of hydrogen and helium – which we are used to encountering as lightweight gases. Under the extreme conditions at Jupiter's centre though, those gases behave like exceptionally dense solids. In fact their density becomes equal to that of the weightiest substance (osmium) that we currently know about – and that is quite a bit heftier than solid gold. At Jupiter's cloud tops, by contrast, the pressure in Jupiter's atmosphere is similar to that of our own.

Jupiter's orbit is slightly elliptical so though its distance from the Sun is usually quoted as 5.20 AU, it actually varies from 4.95 to 5.45 AU. Over Jupiter's 11.86-year orbit, the energy that it receives from the Sun thus varies by ±10%.

If we take hot Jupiters to have masses between a half and twice that of Jupiter and to be less than around 0.5 AU away from their host stars then they comprise about 16% of the currently known exoplanets. Their average mass is close to that of Jupiter, their average distance from their host stars is 0.086 AU and their orbits are about as elliptical as that of Jupiter on average. The host stars' masses average 1.06 times that of the Sun and their radii average 1.17 solar radii. Hot Jupiters are thought to have originated further out from their stars than their present positions and to have migrated inwards at a later date so it is likely that the composition and structure of a newly-formed hot Jupiter would have been very similar to the composition and structure of the newly-formed Jupiter.

Thus the typical hot Jupiter may be imagined to be just as though we moved Jupiter from its present orbit to one a fifth of the distance of Mercury away from the Sun (13 million kilometres). There are three main effects that would result from such a move. The first and the most obvious effect is that Jupiter in its new position would receive around 3,500 times more energy from the Sun than it does at present. Other things being equal this would then mean that the surface temperature would rise from the present −140°C to about +760°C. In fact, since the orbit is slightly elliptical, then if the planet remained rotating rapidly the temperature would vary from +740°C when furthest from the Sun (aphelion) to +790°C when closest to the Sun (perihelion). If, as would be quite probable, tides caused the planet's rotation to become locked onto the Sun so that it always kept the same face towards the Sun, then the temperature at the sub-solar point could rise to ~1,000°C whilst falling to ~600°C or so on the side away from the Sun. This would lead to 5,000–10,000 km/h winds howling around the planet such as those seen on HD 189733 b (Figure 7.5). The present wind speeds on Jupiter for comparison reach some 300 km/h at maximum.

Despite the enormous change in the energy being received from the Sun, there would be little change as a result of that deep inside the planet – the outer layers of the planet are an extremely effective insulator. Another effect of the tides though would be to heat the planet's interior directly. The strength of the tide from the Sun would 150% larger at aphelion than at perihelion and this would squeeze and relax (slightly) the whole planet every 9 days (the orbital period). Such changing tidal stresses can release vast amounts of energy – Jupiter's tides within its satellite Io, for example, suffice to melt that satellite's interior completely. For our imaginary hot Jupiter, the tidal heating would raise the temperatures throughout the whole of the planet and add perhaps another 100–200°C to the surface temperatures.

The effects of all these changes upon how we see the planet would be profound. The most obvious one would be that the radius would expand, perhaps to around 1.2–1.4 Jupiter radii (a diameter of ~170,000–~200,000 km) due to the increased temperatures in the upper parts of the atmosphere. The average density would be reduced to a half to three-quarters that of water – although as we have seen (Chap. 2) the effect can be much more extreme, with WASP-17 b having an average density similar to expanded polystyrene foam. The present banded structure of the cloud tops of Jupiter (Figure 12.1) would almost certainly disappear but exactly what would replace it is less easy to decide since quite small changes can have profound meteorological effects. Possibly the increased wind speeds (even if the rotation is not tidally locked) would result in a much more featureless appearance – more like that of the present-day Venus – or conceivably the increased energy input would induce turbulence and convection producing Coriolis storms that would make even the great red spot seem trivial.

Perhaps surprisingly, little of the atmosphere of a hot Jupiter would evaporate away since even at over 1,000°C the planet's gravitational field is able to hang on to its hydrogen and helium. However, such a loss of atmosphere *would* start to occur for planets less than around 0.02 AU away from their host stars, and at less than 0.015 AU (2,250,000 km) out the planet would soon lose all its lighter elements leaving just the rocky core (a chthonian planet). Even this mass loss though would not be due to evaporation, but would arise when the outer parts of the planet's atmosphere expanded until they left the gravitational sphere of influence of the planet and moved into that of the star. In some cases the outer layers of the planet's atmosphere may be lost as an intense stellar wind drags them into space. This seems to be the case for HD 209458 b where recent spectroscopic observations made by the HST have shown that the lost material is trailing away from the planet giving it a comet-like appearance. The exoplanet that is closest to its host star found so far (SWEEPS-10 with a separation of 0.008 AU is unconfirmed – see below) is GJ 1214 b with an orbital radius of 0.014 AU (two million kilometres). However although GJ 1214 b is undoubtedly hot, its mass is 0.02 Jupiter masses, making it a super-Earth not a hot Jupiter.

Any satellites, especially the outermost, weakly held ones, would probably have been lost as the planet moved in towards the star, but if not, then they soon would be become detached from the gravitational field of the planet to become small independent planets themselves. Their interactions with the star and planet would then quickly send them crashing into the star or the planet or zooming off into the outer reaches of the planetary system. It is thus likely that any natural satellites of hot Jupiters will have been lost during the planets' migrations inwards towards their host stars. Confirmation of this comes from a recent investigation of 72 transiting exoplanets which failed to find any evidence of satellites.

We may though set limits upon the range of stable orbits that any surviving satellites could adopt. If the satellite is too close to its planet then it will be broken up by tides (the Roche limit). Whilst if the satellite is too far from its planet it will be lost to the host star (the Hill limit).

Taking a 1 Jupiter mass exoplanet in a circular 0.1 AU orbit around a 1 solar mass star as an example, then as shown in Figure 12.2, any natural satellites larger than a few kilometres across must orbit between 0.0013 and 0.0069 AU (200,000 to 600,000 km – 50% to 150% of the separation of the Earth and Moon) out from their exoplanet. Thus if hot Jupiters do still posses any satellites they must be relatively close to the planet. For hot super-Jupiters (see below) the limits roughly double to between 0.0027 and 0.015 AU for a 10 Jupiter mass planet 0.1 AU out from a 1 solar mass host star.

The rise in internal temperature would have relatively little effect upon the internal structure by itself. The effect of the tidal variations though would be to stir up the interior and make its composition much more uniform – whether this would affect the external appearance or not though is less clear.

The recently discovered exoplanet CoRoT-12 b approximately fits the description of a hot Jupiter that we have just deduced by imagining Jupiter being moved inwards from its present orbit

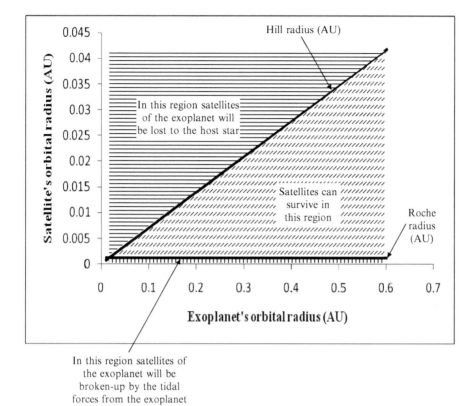

In this region satellites of
the exoplanet will be
broken-up by the tidal
forces from the exoplanet

FIGURE 12.2 The region wherein a natural satellite can survive for a one Jupiter mass exoplanet orbiting a one solar mass host star (the satellite's density is assumed to equal that of water – 1,000 kg/m³).

(see below). Also listed below are details of a selected number of hot Jupiters showing the range of their properties.

Observed data for selected hot Jupiters and their host stars (The notable feature(s) used for the selection is underlined and in bold. Largely based on data from the Extrasolar Planets Encyclopaedia (http://exoplanet.eu/))

Star	Exoplanet mass (Jupiter masses) (M sin i)	Exoplanet radius (Jupiter radii)	Exoplanet average density (× water)	Exoplanet surface temperature (°C)	Orbit period (days)	Orbit radius (astronomical units)	Orbit ellipticity (relative to Jupiter's)	Star mass (solar masses)	Star radius (solar radii)
CoRoT-12 b	0.92	1.44	0.45	1,000–1,500	2.83	0.008	×1.4	1.1	1.1
WASP-6 b	**0.5**	1.2	0.38	1,000	3.36	0.042	×1.1	?	?
HAT-P-15 b	**1.9**	1.07	2.0	?	10.9	0.1	×4	1.0	1.1
Lupus-Tr-3 b	0.8	**0.89**	1.5	?	3.9	0.046	0	0.87	0.82
WASP-17 b (retrograde orbit)	0.5	**1.74**	**0.1**	?	3.7	0.04	×2.7	1.2	1.4
CoRoT-13 b	1.3	0.885	**2.4**	?	4.0	0.05	0	1.1	1.0
SWEEPS-10 (unconfirmed)	?	1.24	?	?	**0.424**	**0.008**	?	0.44	0.41
OGLE-Tr-56 b	1.3	1.2	1.0	1,700	**1.2**	**0.023**	0	1.2	1.3
HD 52265 b	1.1	?	?	?	**119**	**0.49**	×6	1.2	1.25
HD 74156 b	1.88	?	?	?	52	0.29	**×13**	1.2	1.6
HD 27894 b	0.62	?	?	?	18	0.12	×1	**0.75**	0.8
HAT-P-7 b	1.8	1.4	0.85	?2,500	2.2	0.038	0	**1.5**	1.84
HD 192263 b	0.72	?	?	?	24	0.15	0	0.81	**0.75**
HD 185269 b	0.94	?	?	?	6.8	0.077	×6	1.28	**1.88**

One hot Jupiter, while not in itself particularly unusual, is notable as the only exoplanet detected so far that is from another galaxy. The original galaxy would have been a dwarf galaxy orbiting the Milky Way. The tides from our galaxy ripped it apart thousands of millions of years ago until it formed a long stream of ten to a hundred million stars looping several times around the Milky Way. The stars in the stream are all old and have low abundances of elements heavier than helium so that they can be identified separately from the normal stars belonging to the Milky Way. The exoplanet's host star, HIP 13044, is a part of this Helmi stream and lies about 2,000 light years away from us within the southern constellation of Fornax. The planet, discovered by the radial velocity method, has a mass 25% larger than that of Jupiter and orbits just a few million kilometres out from its star every 16 days. The exoplanet just escaped being engulfed by its host star when the latter became a red giant but may suffer that fate a few million years from now when the star expands again.

Super-Jupiters

The division between exoplanets and brown dwarfs is taken to be around thirteen Jupiter masses, so that beyond the hot Jupiters we have a group of some very massive planets to consider. These ultra-massive planets are called super-Jupiters or mega-planets. Now we might well expect that as the mass of the planet rose above two Jupiter masses, so the radius of the planet would rise above the one or so Jupiter radii that we have seen for the hot Jupiters. However that turns out not to be the case. In fact, the radius remains close to around one Jupiter radius for all the more massive exoplanets and all the way up to the top end of the brown dwarfs at some 80 Jupiter masses (Figure 12.3). The reason for this is that as the mass increases, so does the central pressure within the planet. The increased pressure causes the core and the outer central region to compress and collapse thus reducing the volume by at least as much and perhaps by more than the increase in volume due to the extra material. All planets and brown dwarfs from about the mass of Saturn (0.3 Jupiter masses) until we get to the smallest stars therefore have radii between about 100,000 and 200,000 km. In fact the maximum size occurs for hot Jupiters with about Jupiter's mass.

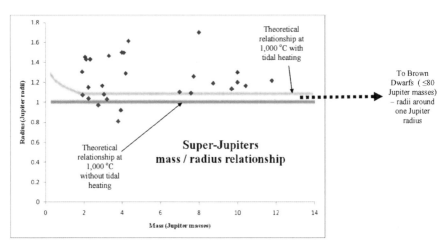

FIGURE 12.3 The observed and theoretical relationships between the radii of large exoplanets (and brown dwarfs) and their masses – the blue diamonds are observed examples.

Of some 180 currently known super-Jupiters, 40 are within 0.5 AU of their host stars and so may be expected to share many of properties of the hot Jupiters. Beyond that, it is uncertain how a hot super-Jupiter might appear to a nearby observer. Since these planets' radii are more-or-less constant, the gravity at their visible surfaces will increase directly with the mass – ten times that of Jupiter for a ten Jupiter mass super-Jupiter and so on. The increase in the gravitational field throughout the planet as well as at the visible surface must surely have some effects upon the visual appearance of the planet, but predicting what those effects might be is beyond the capabilities of present-day meteorology.

Warm and Cold Jupiters and Super-Jupiters

Seventy-eight percent of all currently known exoplanets have masses estimated at 0.5 Jupiter masses or above. Of these, as we have seen, about 26% are hot Jupiters or hot super-Jupiters. Thus around half of all known exoplanets have masses above 0.5 Jupiter masses and are in orbits that have radii larger than 0.5 AU. Half of these exoplanets (25% of the total), which we may call warm Jupiters or warm super-Jupiters, are in orbits between 0.5 and 2

AU and the remainder (cold Jupiters or cold super-Jupiters) have orbits larger than 2 AU. The cold Jupiters and super-Jupiters may be expected to resemble our own Jupiter in many respects – perhaps with ring systems like Saturn's and with numerous satellites, since these will not have been lost during migrations into much smaller orbits. One of the closest matches to Jupiter itself is 23 Lib c discovered from radial velocity variations by Hugh Jones *et al* in 2009 using observations made with the AAT (although the complete orbit has yet to be tracked). The host star of 23 Lib c is just slightly cooler than the Sun (spectral type G5V – see Appendix IV for a brief summary of stellar spectral and luminosity classification), the exoplanet's mass is 0.82 Jupiter masses and it is in a nearly circular orbit with a radius of 5.8 AU and a period between 12 and 15 years (cf. Jupiter: 5.2 AU and 11.8 years). The warm Jupiters and super-Jupiters will have migrated inwards to some extent and so will probably have lost any ring systems and some or all of their satellites. In other respects we may expect them to be intermediate between the hot and cold Jupiters and super-Jupiters.

Eccentric Jupiters

A sub-group of the massive exoplanets that overlaps with those just considered is the eccentric Jupiters. These are Jupiter-mass or greater exoplanets that are in very elliptical orbits. About 28% of all known exoplanets have masses above 0.5 Jupiter masses and are in orbits that are between twice and ten times more elliptical than that of Jupiter. Another 8% have orbits more than ten times as elliptical as Jupiter's. The most extreme example found to date of an eccentric Jupiter is HD 20782 b. This is an exoplanet with a mass of at least 1.8 Jupiter masses in an orbit that is 20 times more elliptical than that of Jupiter around a host star that is very similar to the Sun. At its closest point to the star (periastron) HD 20782 b is just 0.1 AU away from its star and at its furthest (apastron) it is 2.62 AU out. The energy that the exoplanet receives from its star thus varies from 4 to 2,700 times that received by Jupiter. Furthermore the internal tidal heating must be prodigious. Quite what the resulting planet must look like is currently impossible to guess although something pretty weird and possibly quite spectacular would seem likely.

Hot and Cold Neptunes

Exoplanets with masses more than about ten Earth masses (0.03 Jupiter masses) but significantly less than that of Jupiter are labelled as hot or cold Neptunes (Neptune's mass is 17 Earth masses). About 17% of currently known exoplanets fall into these categories although the mass range from ~100 to ~150 Earth masses (~0.3 to ~0.5 Jupiter masses – 6% of exoplanets) should perhaps be considered to be a transition zone from the Jovian mass exoplanets. Hot Neptunes (orbit radius <0.5 AU) comprise 13% of all known exoplanets and warm Neptunes (0.5 < orbit radius < 2 AU), 1.5%. No exoplanets are known to be as far out from their host stars as Uranus and Neptune are from the Sun (19 and 30 AU respectively) and only six are beyond 2 AU. OGLE-06-109-L c and OGLE-2007-BLG-368L b though, both recently discovered by microlensing, have masses of ≥90 Earth masses and ≥23 Earth masses combined with distances from their host stars of ~4.5 and ~3.3 AU respectively. Both host stars are small and dim with luminosities between 10% and 20% that of the Sun. In the solar system therefore, these two planets would be at equivalent distances of 12 and 9 AU and so are likely to be very similar to Uranus and Neptune (Figure 3.4).

Super-Earths

The last group of main-stream exoplanets for which we have observed examples is the super-Earths. A super-Earth has a mass between ~2 and 10 Earth masses, but may not have much else in common with the Earth. Because the usage of 'Earth' in this context tends to be taken to imply an Earth-like planet, the terms super-Venus, super-Pluto and Gas Dwarf are sometimes used as alternatives. Twenty-four super-Earths (5% of all exoplanets) are currently known (pulsar planets are considered separately). Gliese 581 e has the lowest mass of any known exoplanet orbiting a normal star at 2 Earth masses, but it is just 0.03 AU out from its host star. That star is considerably cooler and fainter than the Sun but nonetheless Gliese 581 e most probably resembles a large version of Mercury (Figure 12.4). Tidal heating of the planet's interior may have led though to a geologically active surface including volcanoes.

FIGURE 12.4 An artist's impression of Gliese 581 e (foreground) together with its host star and two of the other five exoplanets belonging to the system. (Reproduced by kind permission of ESO and L. Calçada).

Gliese 876 d (one of four exoplanets belonging to Gliese 876 – a cool red dwarf star 15 light years away from us in Aquarius) has a mass of 6.7 Earth masses and is to be found just 0.02 AU (3 million kilometres) from its host star. The host star however has a luminosity only 0.2% that of the Sun. Gliese 876 b thus experiences about the same irradiation as a planet about halfway between Mercury and Venus would within the solar system.

None of the known super-Earths have any real resemblance to our own Earth, but overall, OGLE-05-390L b has the most in common. This is a 5.6 Earth mass exoplanet in a 10-year orbit 2.1 AU out from a cool red star. The planet may have a thin atmosphere, but most volatile substances such as ammonia, methane and nitrogen will have frozen out at its probable temperature of –220°C. It could be a rocky planet or resemble a small version of Uranus. Almost all the other super-Earths are close in to their host stars and will be very hot like Gliese 581 e. The overall distribution of cold, warm and hot exoplanets with their mass group is shown in Figure 12.5.

The smallest exoplanet of all is the innermost one of the three belonging to the pulsar PSR 1257+12. Its mass is 0.02 Earth masses

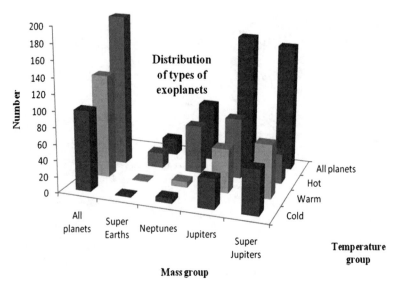

FIGURE 12.5 Distribution of exoplanet types.

(a bit less than twice the mass of the Earth's Moon), the system's other two exoplanets have masses of 4.3 and 4.0 Earth masses. The distances of the exoplanets from the pulsar are 0.19, 0.36 and 0.46 AU respectively. As discussed earlier there are two possible ways in which these planets might have originated – either as the rocky cores of gas-giant planets whose outer layers were stripped away during the supernova explosion that produced the pulsar or formed from the debris left behind after the explosion. Howsoever these exoplanets may have come in to existence, they are unlikely to resemble any of the other known exoplanets. Their temperatures are probably very low, since the total luminosity of the pulsar is small, but they are likely to be bathed in intense particle bombardments and, of course, if caught in the pulsar's beamed emissions, subject to quite unimaginable levels of radio, x-ray and gamma-ray pulses 160 times every second.

Free-Floating Exoplanets

Free-floating planets are not considered by some workers to be 'proper' exoplanets. However, as discussed at the start of this book, they are included here, if only because at least some must have originated as 'proper' exoplanets within an exoplanetary system and only later flung out to take up an independent existence following gravitational interactions with other planets within that system. The majority of free-floating planets though almost certainly formed directly by condensation out of an inter-stellar gas and dust cloud in a similar fashion to the way in which stars and brown dwarfs are produced.

Although such newly formed free-floating exoplanets clearly have no significant input of energy from any external source they will, for a few million, perhaps a few tens of millions of years, have energy from their initial collapse being released internally. If they are blanketed in a thick insulating atmosphere, the temperature at any solid surface could be quite high – from perhaps –100 to 1,000°C. That temperature range includes the 0–40°C that we find comfortable and so, improbable as it may seem, they could be inhabitable. Free-floating planets seem unlikely to develop life of their own in any way resembling terrestrial life however because of the absence of visible light (for photo-synthesis) and the relatively brief interval before they cool down to the temperature of inter-stellar space. In the absence of an insulating atmosphere, free-floating planets will quickly cool to –260°C or so and all free-floating planets will do so in 100 million to 1,000 million years. Currently known free-floating planets have masses ranging from 2 to 10 Jupiter masses and temperatures at their visible surfaces of 1,000–2,000°C – there seems however to be no reason why Earth-sized and smaller free-floating planets should not also exist.

Exoplanetary Systems

Most exoplanet host stars are only known to possess a single planet, though it seems probable that there are more to be found in most such cases. There are however over 50 stars with two or more detected exoplanets. These exoplanetary systems contain about a quarter of all currently known exoplanets and up to seven exoplanets within a single system (although some remain to be confirmed). 55 Cnc A (also called ρ^1 Cnc) has five known exoplanets. HD 10180, a solar-twin star 130 light years away from us in Hydrus, was found in 2010 to have six confirmed exoplanets (Figure 3.17) with a seventh awaiting confirmation. The unconfirmed exoplanet, HD 10180b, could be just 40% more massive than the Earth, but at just over three million kilometres out from its host star its surface temperature is likely to exceed 2,000°C – hot enough that it is probably a completely molten mix of lava and liquid metals. Another six-exoplanet system is the very recently discovered Kepler-11 (Figure 1.1b) and Gliese 581 has four confirmed and two unconfirmed exoplanets (Figure 1.1a).

Large multi-exoplanet systems

Planet	Minimum mass (Jupiter masses)	Orbital period (days)	Orbital radius (astronomical year units)	Discovery	Comment
55 Cnc b	0.83	14.7	0.11	1996	
55 Cnc c	0.17	44	0.24	2002	
55 Cnc d	3.8	5,200	5.8	2002	
55 Cnc e	0.024	2.8	0.038	2004	
	0.033	0.74	0.016		Alternative interpretation of the data for 55 Cnc e
55 Cnc f	0.14	260	0.78	2007	
Gliese 581 b	0.049	5.4	0.041	2005	
Gliese 581 c	0.0169	12.9	0.07	2007	
Gliese 581 d	0.0223	66.8	0.22	2007	
Gliese 581 e	0.0061	3.15	0.03	2009	
Gliese 581 f	0.023?	433?	0.76?	?	Unconfirmed
Gliese 581 g	0.01?	36.7?	0.15?	?	Unconfirmed
HD 10180b	0.004?	1.2?	0.02?	?	Unconfirmed
HD 10180c	0.041	5.8	0.06	2010	
HD 10180d	0.037	16.4	0.13	2010	
HD 10180e	0.079	49.7	0.27	2010	
HD 10180f	0.075	123	0.49	2010	
HD 10180g	0.067	601	1.42	2010	
HD 10180h	0.203?	2,200	3.4	2010	
Kepler-11 b	0.014	10.3	0.09	2011	
Kepler-11 c	0.043	13.0	0.11	2011	
Kepler-11 d	0.019	22.7	0.16	2011	
Kepler-11 e	0.026	32.0	0.19	2011	
Kepler-11 f	0.007	46.7	0.25	2011	
Kepler-11 g	<0.95	118	0.46	2011	

The effect upon an exoplanet of being within an exoplanetary system seems likely to be small. Once the planets have settled down to stable orbits, their appearances, characteristics and structures will be very little different from the case if they were the only exoplanet belonging to that star.

The recently discovered pair of planets orbiting HD 200964 (a cool sub-giant star 200 light years away from us in Equuleus) are in orbits that are probably stable despite their being separated by only 0.35 AU at times. These two planets have minimum masses of 1.85 and 0.9 Jupiter masses and are in 614 day (1.6 AU) and 825 day (1.95 AU), slightly elliptical orbits. At their closest approach to each other their mutual gravitational attractions are four and a half times that which holds the Earth and the Sun together. Normally such a large interaction would quickly change the exoplanets' orbits until they moved further apart, but in this case four times the orbital period of the inner planet is very close to equaling three times the orbital period of the outer planet. Such a 4:3 resonance, as it is called, can be stable, as it is for Saturn's satellites Titan and Hyperion. For HD 200964 b and HD 200964 c the stability remains to be confirmed, but it would seem to be very likely. The regular close approach between the two planets might suggest that if human astronauts were to stand on the surface of HD 200964 c (say – or more plausibly on the surface of one of its hypothetical moons) then HD 200964 b would loom enormously and threateningly in their sky every 6½ years. In fact a quick calculation will show that even at the planets' closest approach to each other, the inner planet would only be about nine minutes-of-arc across when seen by such observers. The most acute human vision can resolve about three minutes-of-arc, so the planet would actually appear to resemble very bright star that might just be seen to have a disk. A similar system announced at the same time as HD200964 is that of 24 Sex. This has two Jovian mass exoplanets in a 2:1 resonance, though they are 0.7 AU apart at their closest.

Perhaps the most similar planetary system to our solar system found to date is that of 61 Vir – although the resemblance is marginal. 61 Vir is a star that is similar to the Sun, but slightly cooler and fainter. It is located about 30 light years away from us and should be visible to the unaided eye from a good observing site. Three exoplanets have been detected in orbit around the star via the radial velocity method and a fourth one is suspected but not confirmed. The innermost planet is a super-Earth (>5 Earth masses) and the two outer planets have masses similar to Uranus (>18 Earth masses and >23 Earth masses). There however the resemblance to the solar system stops – all three planets have orbits smaller than that of Venus. 61 Vir b, the super-Earth, is just 0.05 AU out from the star and its surface temperature is likely to be around 1,000°C. Furthermore 61 Vir b's orbit is quite elliptical so that its distance from the star varies from 6,600,000 to 8,400,000 km.

Exoplanets in Binary or Multiple Star Systems

Of more significance to the nature of an exoplanet than the presence of companion planets may be when the exoplanet belongs to a binary or multiple system of stars. In the movie series 'Star Wars', the home planet of Luke Skywalker orbits an imaginary binary star system comprising a solar-type star and a star somewhat cooler than the Sun. The stars are named Tatoo I and Tatoo II and the planet, Tatooine. Exoplanets belonging to binary and multiple star systems are thus sometimes called Tatooine planets.

Since binary and multiple star systems comprise around half of all stars, exoplanets within such systems are likely to be quite commonplace. It is even possible that Tatooine exoplanets could be more common than those of single stars. It has been estimated that at least half of binary star systems possess regions wherein stable orbits for planets exist and it has also been suggested that planet formation could be enhanced in a binary system because the accretion rate of proto-planets is likely to be improved when the proto-planetary disk of one star is stirred-up by the presence of the second star.

In fact 55 Cnc A and its five planets (above) may form a binary star system with 55 Cnc B which is a 0.13 solar mass red dwarf. However 55 Cnc B is 1,100 AU away from the main star – 200 times the distance of even the outermost exoplanet – so even if the two stars are gravitationally linked there is probably no resulting effect upon the planets. The exoplanets of 55 Cnc A are an example of planets in an S-type orbit within a binary star system. S-type orbits are when the planets orbit around a Single star. Another example of an exoplanet in an S-type orbit that we have already seen is γ Cep A's exoplanet. It is in a 2.48 year orbit, has a minimum mass of 1.7 Jupiter masses and is 2.13 AU out from its host star. γ Cep B, the companion star, is in a 67-year, 20-AU orbit and so is just ten times the distance of the planet from γ Cep A – this is the closest currently known relative proximity of any exoplanet and companion star. Even in this case though, there would be little effect upon the planet arising from the companion star except to give any ET inhabitants of the planet a spectacularly bright star in their sky.

When the two stars are close together and the exoplanet orbits around their common centre of gravity, they are said to be in a P-type orbit. Methuselah (or PSR 1620-26 b) is in such an orbit. Methuselah orbits a neutron star-white dwarf binary pair that are in a mutual 6-months, 0.8 AU orbit around their common centre of gravity. The exoplanet is 23 AU out and has an orbital period of around a century.

Clearly almost any exoplanet discovered by timing an eclipsing binary star's period variations is likely to be in a P-type orbit. Thus another example of a P-type orbit is for the planet of HW Vir. HW Vir comprises two small cool stars 0.005 AU apart in a 2.8-h orbit around each other. The planet has a mass at least 8.5 Jupiter masses and is 3.6 AU out in a 9-year orbit. The situation is complicated however by the presence of a probable 19 Jupiter mass brown dwarf at 5.3 AU from the binary stellar system.

Some binary stars alter their orbits with time. In particular binaries in which both stars are spinning rapidly are likely to have strong magnetic fields and intense stellar winds. The stellar winds slow the stars' orbital motions causing them to move into tighter (and actually faster) orbits. The stars' orbital changes are likely to affect any exoplanets that they may possess drastically. Evidence for such a process has recently been detected by NASA's Spitzer spacecraft. Spitzer observed three close binary stars, known as RS CVn variables, and found that they were surrounded by dusty disks. Since these are mature stars, they should long ago have lost any material left over from their origin. These disks must therefore have formed recently. It is suggested that the disks may have originated from colliding and disintegrating exoplanets whose orbits were changed by the changing orbits of the stars.

In 2005 a Jupiter-mass exoplanet was reported orbiting the star HD 188753 every 3.3 days. The discovery has been disputed and is still regarded as unconfirmed. If the planet does exist, then it will have a curious life. The host star is a part of a triple stellar system. HD 188753A is a solar-type star and has a binary star composed of spectral type K and M dwarf stars orbiting it every 25 years. The two cooler stars orbit each other in a period of 156 days and are about 0.7 AU apart. The orbit of HD 188753B and C around the primary star has a radius of 12.3 AU but is highly elliptical At one point therefore the binary stars are about 6 AU from the primary and are 18 AU away at the opposite end of the orbit. About four times a century therefore, the exoplanet (if any) which is about 0.04 AU away from HD 188753A will undergo a close approach from the binary star pair (Figure 12.6). Whether this is a

FIGURE 12.6 Artist's concept of the HD 188753 system viewed from an hypothetical small rocky planet and with the Tatooine exoplanet HD 188753 b imagined to have a Saturnian-type ring system and satellites. (Copyright © C. R. Kitchin 2010).

stable situation or not and quite what the binary star passage will do to the planet through gravitational perturbations and energy input has yet to be determined, but it seems likely that the planet will fulfill the ancient Chinese curse of 'may you live in interesting times' to the full. The presence of the binary stars in the system when the exoplanet was forming would probably have prevented that formation. It seems likely therefore that the exoplanet first formed around HD 188753A and only later, after the planet had migrated inwards towards its host star, were the binary stars also captured to form the triple system.

ψ^1 Aquarii is a fourth magnitude orange giant star some 150 light years away from us. Its exoplanet, a 2.9 Jupiter mass hot Jupiter may hold the record for the exoplanet with the most host stars. ψ^1 Aquarii is part of a triple system with two dwarf stars. The dwarf stars form a binary with a separation of 18 AU and lie 2,250 AU away from the main star. However another nearby binary star system may also be gravitationally linked to the triple – making five host stars for one exoplanet.

Host Stars and Their Effects upon Their Exoplanets

An obvious major influence upon the nature of an exoplanet is the luminosity and temperature of its host star. These are usually specified by the star's spectral and luminosity classes (see Appendix IV for a brief summary of stellar spectral and luminosity classification). The luminosity of the host star has the principal effect upon the nature of any exoplanets, while the star's surface temperature changes the balance between the amount of ultra-violet radiation emitted compared with the visible and infrared radiation.

The most luminous exoplanet host star yet found is γ^1 Leonis which is around two hundred times brighter than the Sun. γ^1 Leonis b is a >8.8 Jupiter mass exoplanet but it is 1.2 AU out from its host so despite the brightness of its star, the exoplanet is probably a relatively conventional hot Jupiter but with reduced internal tidal heating.

The faintest host star is SCR 1845, a faint red dwarf star 13 light years away in the constellation of Pavo. Its mass is estimated to be just on the star/brown dwarf boundary (70–80 Jupiter masses), so it is possible that this may not be a true star after all. Nonetheless, SCR 1845 is just 0.0001 % of the brightness of the Sun, so SCR 1845 b, its >8.5 Jupiter mass exoplanet, at 4.1 AU out from the star receives the same energy as a planet 90 times further away from the Sun than Pluto would do. Effectively the planet receives no energy input from its host star and so must resemble free-floating planets in its characteristics and appearance. 2M1207 with a mass of 25 Jupiter masses is definitely a brown dwarf and it acts as a host to a 3–10 Jupiter mass exoplanet. Although 2M1207 is around 15 times brighter than SCR 1845, its planet is about 50 AU away from it. 2M1207 b's illumination is thus only 15 % that of SCR 1845 b and so it too must resemble a free-floating exoplanet. In fact the 2M1207 system is thought to be very young, so the exoplanet is still glowing from the heat generated during its formation and can be seen directly (Figure 3.14).

The most brightly illuminated exoplanet found to date is probably WASP-33 b which is 400 light years away from us in Andromeda. Its host star has a likely spectral class of A5 and the planet is in a 0.026 AU orbit. The four Jupiter mass exoplanet therefore receives over 200,000 times more energy per square metre than does our own Jupiter – an incredible 12 mega-watts per square metre. For comparison, when the Sun is overhead on the Earth, we receive about 1.5 kW per square metre on the Earth, providing that the sky is clear. A large power station on the Earth typically generates 1,000 mega-watts, so an area of WASP-33 b's surface just 10 m × 10 m (33 ft × 33 ft) receives the equivalent of the entire output of a large terrestrial power station. The temperature at the cloud tops of WASP-33 b is likely to be in the region of 3,000°C – hotter than some of the coolest stars and certainly hotter than many brown dwarfs (the brown dwarf SSDS1416 + 13B, for example, has an estimated surface temperature of about 500 K).

The distribution of exoplanet host stars with star type/luminosity class, where known, is shown in Figure 12.7. The distribution mirrors quite closely that of stars in general except that subgiant exoplanetary hosts are about three times commoner than might be expected. This, though, is probably an observational bias reflecting the greater distances to which subgiants can easily be

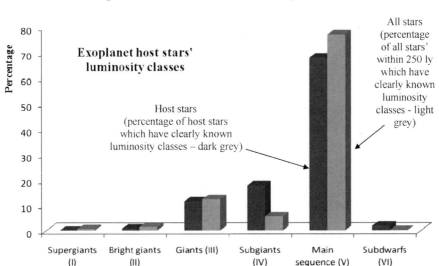

FIGURE 12.7 Distribution of exoplanetary host stars' luminosity classes (as a percentage of those host stars whose luminosity class is clearly known) compared with the distribution of luminosity classes amongst stars within 250 light years of the Earth.

studied due to their greater luminosities when compared with main sequence stars. The volume of space so far sampled for subgiant hosts is thus larger than the volume of space sampled for main sequence hosts – once sufficient exoplanets are known for the data for all type of host stars to be limited to a certain distance (as has been done for the 'all stars' data shown in Figure 12.7), this anomaly will probably disappear. The two objects known to orbit bright giants are probably small brown dwarfs, but since some uncertainty remains, have been included in Figure 12.7.

A few exoplanets orbit pulsars or binary systems containing white dwarfs – PSR 1257+12 a, b and c, Methuselah and DP Leonis b for example – but the numbers involved are small and the exoplanets likely to be rather different from 'conventional' exoplanets, as discussed earlier.

Host stars' temperatures determine their spectral classes and the hottest stars known to possess exoplanets are of class B (Figure 12.8). If these were main sequence stars (luminosity class V) then their luminosities would be up to 3,500 times brighter than the Sun. However all examples of such hot exoplanet hosts currently

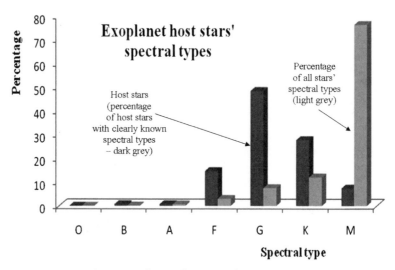

FIGURE 12.8 Distribution of exoplanetary host stars' spectral types (as a percentage of those host stars whose spectral type is clearly known) compared with the distribution of spectral types amongst all stars.

known are subdwarf stars (class VI) and so are typically 'only' 50 to 100 times the solar brightness. Thus V 391 Peg b, which is a 3.2 Jupiter mass exoplanet orbiting 1.7 AU out from a subdwarf B class star, receives about the same energy per unit area as we do on the Earth and would be classed as a warm Jupiter or warm super Jupiter. Where a difference may arise is that the host star emits some 50% of its energy as ultra-violet and shorter wavelength radiation compared with about 10% for the Sun. The bulk effect on the exoplanet of the different balance of the radiation is not clear, but at a molecular level it is likely that few molecules, such as the methane and ammonia found on Jupiter, will be able to survive near the planet's cloud tops since they will rapidly be disassociated by the intense UV radiation.

The hottest main sequence stars yet found with exoplanets are Fomalhaut (α PsA – spectral type A3 V, temperature 8,300°C) and HR 8799 (spectral type A5 V, temperature 7,200°C). Both these stars and their exoplanets have been imaged directly (Figure 7.3) so the planets are all in orbits quite distant from their stars. Fomalhaut b has a poorly determined mass (0.05–3 Jupiter masses) but its orbital radius is known reasonably precisely to be around 115 AU. Thus despite Fomalhaut being nearly 18 times brighter than the Sun, Fomalhaut b receives less than 4% of the energy from its host star than Jupiter receives from the Sun. HR 8799 is around 15 times brighter than the Sun, but its three planets similarly receive only 2–25% of the intensity of radiation received by Jupiter. Even so, HR 8799 b's and HR 8799 c's surface temperatures seem to be about 900°C and 800°C respectively – presumably largely due to heat left behind from their formation or currently being generated by the continuing contraction of the planet. Although some 20% of both the stars' radiation is in the ultra-violet, their planets are likely to be fairly conventional cold super Jupiters.

Some 20 or so main sequence stars with identical or very similar spectral types (mass 95–105% that of the Sun, radius 95–105% that of the Sun) to that of the Sun possess exoplanets. None of these however has any known exoplanets with masses comparable to that of the Earth. In fact the lowest mass exoplanet in this group is a hot Jupiter orbiting HD 16141, a star in Cetus that is not quite visible to the naked eye. The planet is in an orbit similar to that of that of Mercury and has a mass about a fifth of that of Jupiter. The exoplanet most similar to Jupiter, and probably very close to being a Jupiter-twin, is HD 86226 b, 150 light years away from us in Hydra. It has a mass of 1.5 Jupiter masses and is in a 4.2-year orbit, 2.6 AU out from its solar-twin host star (Jupiter – 11.8 years, 5.2 AU). The remaining exoplanets of solar-type stars are all hot or warm Neptunes, Jupiters or super Jupiters.

The most massive star with an exoplanet found so far, BD20 2457, is 600–700 light years away from us in Leo. Its mass rather uncertainly placed at 2.8 solar masses and it has a ≥12.5 Jupiter mass warm Jupiter (or possibly brown dwarf) in a 620 day, 2 AU orbit. Despite its mass (which would make it a hot spectral class A, verging on a B class, star if it were a main sequence star), because it is a bright giant its spectral type is K and its surface temperature is just 3,800°C, it is still though over 600 times brighter than the Sun. There is also a 21 Jupiter mass brown dwarf in the system orbiting the star every 380 days in a 1.5 AU radius orbit.

Unlike the distribution of host stars with luminosity class (Figure 12.7), their distribution with spectral class, shown in Figure 12.8, differs markedly from that for all stars. While there may well be real differences in the proportion of stars of a particular spectral class that act as hosts to exoplanets, at the moment the disparity seems much more likely to arise from observational biases – O and B class stars are massive, making any radial velocity changes arising from an orbiting exoplanet very small and difficult to detect – K and M class stars are of low luminosity and can only be observed at relatively small distances away from us.

Gas and Dust Disks

Finally some exoplanetary host stars are still embedded within flattened disks of gas and dust. These disks may be the stuff out of which the exoplanets have formed or debris left over from collisions between exoplanets, or both. Howsoever the disks originate, we may still ask 'What would it be like to be on an exoplanet within such a disk?'. The answer may be surprising, given the apparent concentration of material in some disks (see that of β Pic for example – Figures 3.12 and 7.3d). It is that in the short term it would make little difference to the nature of the exoplanet whether the disk were to be present or not. Any ET inhabitants of the exoplanet, though, would notice some minor but striking cosmetic changes to their views of the universe.

The density of the disk material, if it is gaseous, is very low. Precise values for the density are difficult to obtain, but one estimate for the total mass of smallish solid objects (sometimes called planetesimals) within the disk of β Pic is less than that of Jupiter. The mass of the gas could be several tens or hundreds of times this value. Let us, as a basis for calculation, take the mass of the gas in β Pic's disk to be equal to that of the Sun (2,000,000,000,000,000, 000,000,000,000,000 $kg = 2 \times 10^{30}$ kg) and concentrated into a ring doughnut shape (β Pic's planet has cleared the central region) with an inner radius of 35 AU, an outer radius of 200 AU (the actual outer observed edge of the disk is up to 2,000 AU in radius, but the material is centrally concentrated) and a thickness of 20 AU. The mean density of the material in such a disk is about 0.000,000,02% ($= 2 \times 10^{-10}$) of that of the Earth's atmosphere at sea level. Low as this density may appear to us, it is still about 50,000,000,000 ($= 5 \times 10^{10}$) times the density of the material between the solar system's planets. In terms of the number of particles (hydrogen molecules and helium atoms for the disk, nitrogen and oxygen molecules for the Earth's atmosphere), a depth of 10 AU within the disk is equivalent to the vertical depth of the Earth's atmosphere.

β Pic b is well within the central cleared zone of its host star's circumstellar disk since the exoplanet's orbital radius is estimated at between 8 and 15 AU. One of its hypothetical ET inhabitants during the 'day' time would thus see a bluish star about 7% as bright as the Sun and with about 15% of the Sun's angular diameter. The inhabitants' 'nights' however would not be dark, but would be illuminated by a bright blue belt of light covering around half their sky. This is the light from the host star scattered back towards the planet by the circumstellar disk (Figure 12.9). Even in the directions away from the disk, the sky would still have a bluish haze, since the clearance of the central region is only partial – there will still be plenty of gas left to scatter the star's light back towards the planet. It is quite possible therefore that the inhabitants would be unable to see any stars other than their own and would thus know nothing about the existence of the rest of the universe.

The solid components of the disk such as dust particles, pebbles, boulders, asteroids and perhaps small planets will interact with the star's main exoplanet in the same way that such objects

FIGURE 12.9 Artist's impression of the night-time view from an hypothetical satellite of β Pic b. The bright band in the sky is the host star's light scattered back towards the exoplanet by the material in the surrounding disk. (Copyright © C. R. Kitchin 2010).

do within the solar system – i.e. via collisions. An individual collision is unlikely to differ in any significant way from a similar collision within the solar system whether it be a micro-meteorite producing a meteor if it hits an atmosphere or a 0.001 mm 'crater' if it hits a piece of rock or, at the other end of the scale, an asteroid impact that ejects a massive plume of gas if it hits the upper atmosphere of β Pic b or excavates a Mare Imbrium-sized basin from one of the planet's solid satellites.

What will be different from the solar system case though is the frequency of such impacts. The evidence from our own Moon and other solid objects within the solar system that lack an atmosphere, regarding the cratering rate during the first few hundred million years of the life of the solar system, suggests that β Pic b and any satellites could be experiencing between a hundred and ten-thousand times the current rate of impacts on the Earth. Thus an Earth-sized planet in that region could experience a 'dinosaur-killer' impact every million years, a 'Canyon Diablo' (the 1-kilometre-sized impact crater in Arizona) impact every 10 years and a 'Tunguska' (the 30 mega-ton impact of a small comet in Siberia in 1908) impact once a month. Meteor storms, on objects with atmospheres, would be the normal background rate of meteors and any unfortunate astronaut on the exposed surface of an atmosphereless planet or satellite would suffer a bombardment comparable to the worst machine-gun volleys of the First World War.

In the longer term (a few million years) the orbits of β Pic b and any other exoplanets will change as the planets experience drag from the material forming the disk and/or the gravitational perturbations arising from it. In most cases the orbits contract, sending the planets in towards their host star and perhaps changing β Pic b from being the coolish super-Jupiter that it is at present to become a warm or hot super-Jupiter. As previous noted, the planet is likely to lose some or all of its satellites during such an inward migration. Smaller exoplanets than β Pic b are likely to be disturbed by gravitational perturbations from the major planet as well as experiencing drag and perturbations from the disk material. Their orbits could be changed drastically – sending them inwards to collide with the star, causing a collision with β Pic b itself, or putting them into highly elliptical orbits and forcing them to journey many astronomical units into the outer reaches of the planetary system. It is even possible that some of the minor objects could be ejected entirely from β Pic thus becoming free-floating exoplanets or free-floating exo-asteroids.

What else Might Be Out There?

Going beyond the likely characteristics and natures of currently known exoplanets, it is possible to speculate about other types of exoplanet. Since it is clear, even with conservative estimates, that there are many billions, perhaps trillions, of planets in the Milky Way galaxy alone, the wildest of such speculations will probably be proved to be correct somewhere or other. Amongst the more credible suggestions are for Ocean planets and Iron planets.

An ocean planet is just one that is covered or nearly covered in a layer of liquid. While we might expect on the basis of our experience of the Earth that such oceans would be made of water, they could be composed of ammonia, methane or several other common volatile compounds. Ocean planets would be certain to have an atmosphere, if only from the vapour produced by their liquid layer.

That ocean planets are possible may be seen by imagining that the Earth had three times as much surface water as it does at the moment. This is a very minor change – the proportion of the Earth's mass in the form of surface water would just have to change from 0.02% to 0.06%. The ocean depths would then average 11.4 km and only those mountains currently higher than about 7,000 m (23,000 ft) would still rise above the ocean. Nowhere in Africa, Antarctica, Australasia, Europe or North America would remain above water. The highest peaks in the South American Andes *might* just form a few small islands, while only the Himalayas, Hindu Kush and Karakoram ranges would provide any substantial land surfaces. Increase the Earth's surface water by a factor of three-and-a-half (0.07%) though, and even Everest would then be drowned.

Similarly, if Jupiter's satellite, Europa, were imagined to be moved to roughly the Earth's distance from the Sun, its present thin, icy crust would melt and it would be covered by water oceans some 100 km deep. Saturn's satellite, Titan, has methane

FIGURE 12.10 A radar image from NASA's Cassini spacecraft showing lakes of liquid methane on Saturn's largest satellite, Titan. (Reproduced by kind permission of NASA/JPL/USGS).

and perhaps ethane lakes (Figure 12.10) and a moderate increase in the proportion of those compounds that it contains would soon convert those lakes to oceans covering most or all of its surface.

The weather on an ocean planet would probably be quite boring – much of the Earth's more violent meteorology arises from heat imbalances between the land and the sea. For a rotating planet the general circulation of the atmosphere that is driven by temperature differences between the equator and the poles would remain, perhaps producing planet-wide cloud belts and jet streams and an external appearance analogous to that of Jupiter (Figure 12.1). For a planet tidally locked onto its host star, the winds would flow constantly from the planet's sub-stellar point to the opposite side of the planet, perhaps generating cloud patterns like a many rayed sunburst.

Rocky planets close in to their host stars like CoRoT-7 b and Kepler-10 b may have surface temperatures high enough to melt the surface rocks. These planets will then have much in common with 'conventional' ocean planets except that their oceans will be of lava. They are likely to have thin atmospheres of vapourized rock. Their rotations are also likely to be tidally locked onto their host stars so that their day sides could reach temperatures in excess of 2,500°C whilst their night sides might plunge to less than –200°C. The lava oceans would probably spread some distance around into the cold side of the planet before solidifying so that a levee would build up encircling the coldest parts of the dark side – rather like a monk's tonsure. The term 'Lava Ocean Planet' has been proposed for these objects, but because of their probable surface structure, the author here proposes the term 'Tonsure Planets' instead.

'Iron' exoplanets (also called Cannonball Planets) are likely to be found because the core of the Earth is largely a mixture of solid and liquid iron and nickel. Iron meteorites, thought to have come from the broken-up core of an asteroid (or asteroids), show that this is not an unusual situation. Were an Earth-like exoplanet to lose its outer rocky layers, perhaps in a collision with another similar-sized planet, then the iron core could remain. Gas giant planets are also thought to have rocky and iron cores, and these planets could lose their volatile-element outer layers to become chthonian planets whilst migrating inwards towards their host stars or during the post-main sequence evolution of the host star. It is unlikely that an iron planet would be composed purely of iron (or iron and nickel) but would retain some smallish proportion of rocky and volatile elements and compounds. Since iron and nickel are the densest of the common elements they would sink to form the interior of the planet leaving a surface crust and perhaps an atmosphere formed from the remaining material. An iron planet would probably thus look quite 'normal' to an external observer. It would however have a much more intense gravitational field than its size might suggest – an Earth-sized planet largely composed of iron would have a surface gravity about twice that of the Earth.

The stronger gravitational field would exacerbate the buoyancy effects arising from temperature differences in the atmosphere leading (all other things being equal) to similar but more violent weather patterns than those currently experienced by the Earth. On the solid surface of an iron planet the mountains and valleys are likely to be low and shallow since the higher gravity will make the crust less able to support the weight of a mountain or the buoyancy of a valley.

Rather more fanciful are suggestions for diamond and helium planets. The former (also called carbon or carbide planets) might originate from a carbon-rich proto-planetary disk. The planet would probably have an iron core analogous to that of the Earth, but then have a mantle comprising layers of carbon compounds, graphite and diamond. The crust could be formed from hydrocarbons and any atmosphere from methane and carbon dioxide. Observational support for the possible existence of diamond planets came at the end of 2010. WASP-12 b, a hot Jupiter exoplanet orbiting a solar-type star 900 light years away from us in Auriga, was shown by the Spitzer spacecraft to have an atmosphere whose composition is dominated by carbon-containing molecules.

Helium exoplanets could be the remnants of helium-rich white dwarf stars that have lost a large amount of their mass to companion white dwarfs, neutron stars or black holes until they have shrunk to planetary-sized masses. Even on the very generous definition adopted in this book for exoplanets however, such an object would not be a planet since it has at one time generated its energy through nuclear reactions. Conjectures going beyond diamond and helium planets currently belong to the realms of science fiction – though in exoplanet studies science fiction can often rapidly become science fact.

13. Exoplanets and Exoplanetary Systems: Pasts and Futures

There are still some uncertainties over the details of how planets, brown dwarfs and stars come into being. The basic process however is clear – a diffuse cloud of gas and dust in inter-stellar space whose mass can range from one to ten million solar masses and whose diameter can range from a light year to hundreds of light years collapses under its own gravitational pull and ends up as one, a few, hundreds or thousands of smaller objects whose masses range from a tiny fraction to a 100 or so times that of the Sun. The smaller objects have densities ranging from a tenth to ten times that of water – a factor of some 100,000,000,000,000,000,000,000 (10^{20}) or so times denser than the original inter-stellar cloud.

Exoplanets (and asteroids, satellites, comets, etc.) form during the production of stars from inter-stellar gas clouds, so first we look at how stars are (probably) born. Many astronomy books are profusely illustrated with images of inter-stellar gas clouds because such nebulae can be incredibly beautiful (Figure 13.1). The vast majority of those nebulae though are *not* where stars are being born. Some, like the Keyhole nebula (Figure 13.1a) are places where stars were being born a few millions or tens of millions of years ago. Others, like IC 4406 (Figure 13.1b) occur at the ends of stars' lives rather than at their beginnings. These nebulae and others like them are hot: 10,000°C or more. The gas clouds which are the birth places of stars are cold: –260°C or less.

The gas clouds that are stellar wombs are called Giant Molecular Clouds (GMCs) and they are the largest single structures found within the galaxy – yet their existence was unknown until the 1970s. The elusiveness of GMCs arises from their coldness – they are so cold that their energy is almost all emitted at microwave and radio wavelengths. Thus not until radio detectors, radio telescopes

C. Kitchin, *Exoplanets: Finding, Exploring, and Understanding Alien Worlds*, Astronomers' Universe, DOI 10.1007/978-1-4614-0644-0_13,
© Springer Science+Business Media, LLC 2012

a   b

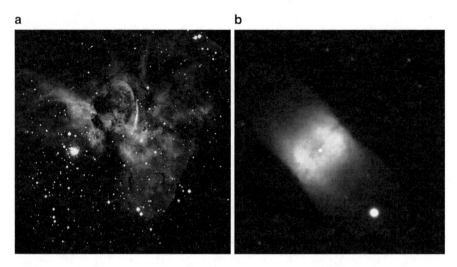

FIGURE 13.1 (a) The Keyhole Nebula in Carina. Imaged by ESO's 3.6-m telescope. (b) The nebula called IC 4,406 in Lupus. Imaged by ESO's VLT. (Reproduced by kind permission of ESO).

and radio astronomy started to come of age could GMCs even be detected, never mind studied in any detail.

The density within a GMC varies from a hundred million atoms and molecules per cubic metre (0.000,000,000,000,000,01 % = 10^{-17}% of the density of the air that you're breathing) to around 10,000 times that value within the small local concentrations called Dense Molecular Cores. As the name of the clouds suggests there are other molecules present within GMCs in addition to the predominating hydrogen molecules (~73% by mass) and helium atoms (~25% by mass). Nearly 300 molecules have now been detected, with the most complex being built up from 70 atoms. Many of the molecules are organic in the chemists' sense of containing one or more carbon atoms. The use of the word 'organic' does *not* imply that the molecules are due to the presence of life forms, although some suggestions have been made that life could originate in inter-stellar space from such molecules.

The nearest GMC is about 1,500 light years away from us in the direction of Orion. It is known as the Orion Molecular Cloud Complex (OMC). It is some 300–500 light years across with a total mass estimated at between a million and ten million solar masses. If we had microwave-sensitive eyes, we would be able to see it as a bright diffuse cloud extending over the whole of the constellation of Orion and into the surrounding constellations. Even without microwave-sensitive eyes though, we can see a very small part of the OMC – the well known Orion nebula (M42) is easily visible to the naked eye as the bottom-most 'star' of Orion's sword. It is one of the closest parts of the OMC to us and is a portion of the cloud where star formation has finished and genuine stars are in existence. Within the Orion nebula, the brightest and hottest of the new stars have heated the gas left behind after their formation until it glows at visible wavelengths.

The coldness of GMCs is the reason why stars form within them. For a gas cloud to collapse, the forces due to gravity pulling inward must be greater than the forces due to the pressure of the gas pushing outwards. The gas pressure depends upon the temperature and density of the gas cloud, while the gravitational force depends upon its mass. For a given gas temperature and density there is a minimum mass, known as the Jeans' mass, that is needed before it will start collapsing. For the conditions inside a dense molecular core within a GMC the Jeans' mass is around one solar mass (Appendix IV).

Thus the start of star formation occurs when a dense molecular core of a GMC collapses under its own gravitational pull. Later in the collapse, as the density of the material rises, the Jeans' mass can fall to 0.001 solar masses or less and Jupiter-mass planets can start to form directly from their own condensations within the overall collapsing proto-star. Studies of all types of companions to stars (exoplanets, brown dwarfs and smaller stars) suggest that many, perhaps most, of the currently known exoplanets may have formed directly in this way.

Planets can also build-up from smaller particles. As the dense molecular core collapses, its density will increase enormously and the dust particles, which form around 1% of the mass of the cloud, will act as nuclei upon which volatile compounds such as water, ammonia, methane, silicates and metals can condense. In the inner part of the cloud where the temperature is higher it will be the silicates and metals that condense. Further away from the centre where the temperatures are lower it will mainly be the ices that will coat the dust particles. In both cases low velocity collisions between dust particles will lead to them sticking together and building up to larger and larger sizes. By continuing the process we may expect gradually to form large dust particles, small pebbles, large pebbles, small boulders, large rocks, small asteroids and eventually bodies with masses comparable to that of the Earth. The Cassini spacecraft has observed the build-up of larger objects from smaller in this fashion within Saturn's rings. The shepherd satellite, Prometheus, stirs up the particles in Saturn's F ring so that they clump together into giant snowballs.

Recent observations of exoplanetary systems and computer simulations suggest that if elements heavier than helium form less than 1% of the mass of the cloud then planetary formation via this process is very slow, while if the abundance is above 1%, planet formation is very much more rapid.

It should be noted though, that while this scenario sounds plausible it has yet to be confirmed by computer simulations, which seem to stall at some point. The collisions then tend to fragment the colliding particles rather than build them up. One possible solution to the problem has been suggested by Peter Goldreich and William Ward and is that the dust in the proto-stellar nebula concentrates towards the central plane of the disk and may become much more dense than the gas. Bodies of kilometre sizes might then be able to be built up rapidly. Towards the end of the process, one of the objects, called the oligarch, starts to dominate its local region. Within that region only the oligarch continues growing. The lesser bodies are either gathered up by the oligarch, collide with each other and are smashed back to fragments or are flung out of the oligarch's region (perhaps then to be gathered up by the next-door oligarch).

Another recent suggestion is that the drag arising from the gas forming the proto-stellar disk will lead to solid objects orbiting within the cloud streamlining after the fashion of racing cyclists. The leading object will clear a path through the gas into which nearby objects will tend to converge since they experience less drag in that orbit. Once a line of objects has formed in this way the leading object will be moving slightly more slowly than the trailing ones because of the drag from the gas. The trailing objects will thus slowly catch up with the leading object and will eventually collide with it at a very low velocity so building up to a larger object rather than breaking up.

While the formation of planets probably mostly happens around young and newly forming stars, it is possible that exoplanets could form around much older stars. In a close binary star system first generation exoplanets might form around one or both stars in the normal fashion during the stars' births. Later when one of the stars starts to evolve towards becoming a giant, material from the expanding star will flow into the gravity well of the other star and form a disk of gas (and perhaps dust) around it. A second phase of planet formation might then be started in this disk of material. A third generation of planets is even possible when the second star starts to evolve and loses material to form a disk around the first star (which is probably now a white dwarf).

Assuming that the build-up of larger objects does continue in some fashion then we will be left with smallish rocky and metallic objects orbiting within the central hotter parts of the proto-stellar gas cloud. Within the solar system we may identify these objects with the terrestrial planets and their satellites and at least some of the asteroids.

Further out within the proto-stellar gas cloud similar rocky and metallic objects are also likely to be the first to form, especially within any condensations undergoing gravitational collapse. However once the temperature falls to less than somewhere in the range −200°C to 0°C – a point within the proto-stellar gas cloud that is called the ice line or the snow line – the water, ammonia, methane and related compounds will start to solidify. These compounds are rich in hydrogen and since hydrogen forms the bulk of the material in the gas cloud, the ices will be abundant. The rocky/metallic cores will thus quickly grow as they accrete the icy material. Eventually the now rocky/metallic/icy bodies will grow large enough for their gravitational fields to attract directly the hydrogen and helium gases forming 98% of the cloud. The growth of the bodies will then become like that of a snowball rolling down a hill and they will quickly build up to the sizes of Saturn, Jupiter, super-Jupiters and even perhaps brown dwarfs.

With the planet formation process just outlined (often called the Solar Nebular Disk Model – SNDM), we should perhaps expect Uranus and Neptune to be even more massive than Jupiter and Saturn. There are several possible explanations for why this is not the case. Firstly the density of the material within the proto-stellar gas cloud will decrease outwards so there may have been less material available for planet formation at Uranus' and Neptune's distances from the Sun. Secondly the formation of the massive Jupiter and to a lesser extent, Saturn, may have disrupted or inhibited continuing planetary formation processes occurring elsewhere. Thirdly the material forming the proto-solar nebula could have been driven away by the formation of the central star. The material could simply have been evaporated when the proto-Sun commenced its nuclear reactions. More probably and in order to explain why the Sun rotates so slowly, it is suggested that the rotational energy of the young Sun was transferred to the remaining gas of the proto-stellar nebula. Such a transfer could occur through turbulence or via magnetic fields. Either way the Sun's rotation would have been slowed down and the remaining gaseous material driven out back into inter-stellar space leaving Uranus and Neptune as poor residues of what they might have been. Yet another possibility is that Uranus and Neptune originated elsewhere in the young solar system and were moved to their current orbits following gravitational interactions with Jupiter and Saturn. This latter process is sometimes called gravitational scattering.

The collapse of a dense molecular core naturally leads to the development of a proto-star that is still surrounded by a considerable amount of material in the form of a disk of gas and dust (generally called the accretion disk or the proto-planetary disk). Some of that material is still falling (accreting) onto the star but much of the rest is available to create planets. The production of exoplanets is thus a natural part of the production of stars – at least for those stars that are single and are of about a solar mass or less. It is less clear that more massive stars will form exoplanets with any frequency since such stars develop rapidly and will quickly become very hot and luminous. The accretion disks of massive stars are thus likely to be driven off before any planets can be produced. Also, while exoplanets are known to exist around binary and multiple star systems, it is likely that the gravitational disturbances arising from the presence of two or more stars during the planetary formation process will disrupt that process in many cases. Nonetheless, the SNDM, if correct, suggests that exoplanets should be extremely abundant throughout the universe.

Slight variations on the process that may have formed the solar system can easily be seen to lead to a wide variety of potential exoplanets and exoplanetary systems. For example, if a single giant planet is formed at an early stage it may inhibit the formation of other planets and leave a system with a single giant planet and perhaps a few Earth-mass and smaller hangers-on. The clearing of the central regions of many of the observed gas and dust disks surrounding young stars by the action of their planets (e.g. β Pic, Figure 7.3) shows how effective this process can be.

Another possibility is if the gaseous material is driven away quickly by the star, then giant planets may have no chance to form and the system will be left just with the smaller planets. Yet another possibility is if the proto-star does not lose its rotational energy, but continues to collapse and fissions into a close binary star system. This could leave several massive planets and doubtless many smaller ones orbiting the stars – for example like the system HW Vir.

What is reasonably clear is that if the process outlined is correct, or even approximately so, it will usually result in the formation of many exoplanets and smaller bodies. The Jupiter and super-Jupiter exoplanets found so far are thus likely to be just the largest of numerous exoplanets forming exoplanetary systems around their host stars.

The planetary formation process just outlined however does *not* produce the hot (or even warm) Jupiter type systems that currently predominate amongst the known exoplanets. There must therefore be an additional process taking place if a newly formed exoplanetary system containing one or more cold Jupiters is to be transformed into one with hot Jupiters. That process is a change in the orbits of the cold Jupiters so that they move in towards their host stars – a phenomenon called orbit migration.

Orbit migration is not a well understood process although it must occur since the inner parts of proto-planetary disks, where hot Jupiters now have their orbits, would be too hot for the volatile materials that largely constitute those gas giant planets to condense. We have already encountered orbit migration amongst rapidly rotating binary stars. For exoplanets four processes may lead to similar results. Type I migration occurs for planets still embedded within gaseous and dusty material. The planet's gravitational field sets up disturbances (spiral density waves) within the surrounding material. These disturbances in turn interact with the planet. The effects of the disturbances outside the planet's orbit are larger than the effects of those inside the orbit. The imbalance between the inner and outer forces causes the planet's orbit to contract and the planet moves in towards its host star. Type II orbit migration arises when a larger planet (at least 10 Earth masses) has cleared its neighbourhood of material. The outer proto-planetary disk however still exists and material from that disk continues to push into the cleared zone. This forces the gap and its planet closer to the host star. It is this process that is likely to produce hot Jupiters.

The third process altering orbits has already been encountered several times and is sometimes given the name of gravitational scattering. It is the perturbation of one planet's orbit by the gravity of another planet. When one planet is much more massive than the other – say perhaps when a potential hot Jupiter is migrating towards its final destination through a region of terrestrial-type planets – then the less massive planet's orbit can be drastically altered. That planet may be flung completely out of the planetary system, into its outer reaches or inwards towards the star. It could also be forced into a collision with the larger planet or other planets or with the star. Unless there is a collision, the larger planet would be relatively little affected by the encounter. When the planets are of comparable masses, then both orbits will be changed, but probably only by small amounts unless the planets pass very close to each other.

The final type of orbit migration suggests that many hot Jupiters are likely to be relatively short-lived planets. The planet raises a tidal bulge in the material of the star similar to the tidal bulge raised in the Earth by the Moon. When the star rotates more slowly than the planet moves around its orbit, that tidal bulge lags behind the planet and acts as a brake to try and slow it down. In fact what happens is that the planet moves into an orbit closer to the star and its orbital motion actually speeds up – thus exacerbating the difference with the star's rotation. This process will continue until the planet crosses the Roche limit and is torn apart by tidal stresses. Ultimately the debris from the planet is likely to crash into the star. The orbital period of the exoplanet OGLE-TR–113 b, discovered in 2002, is currently decreasing by 60 ms every year.

Since the planet's period is just 32 h, if the decrease continues at the same rate, then the planet will crash into its host star in about two million years from now.

Data from the 1,600 or so exoplanetary candidates detected by the Kepler spacecraft to date suggest that the exoplanets closer to their host stars than about 0.1 AU are fewer than expected. This could be evidence for the inward migration and break-up of exoplanets process just discussed. However if the star's rotational period is faster than the planet's orbital period then the tidal bulges lead the planet and can result in the transfer of rotational energy from the star to the planet. This would halt or even reverse the inward migration of the planet and so would also reduce the number of very close-in exoplanets. Quite probably both processes contribute to the lack of exoplanets near their host stars. (Any readers concerned that the tidal drag process might lead to the Moon crashing into the Earth can sleep easily – the Earth rotates faster than the orbital motion of the Moon and so tidal drag is in fact slowly driving the Moon further away from the Earth).

Within a few million years of its formation, the proto-star will lose its surrounding nebula. Several processes may act to remove the material forming the proto-stellar disk – the material may be accreted by the star or by planets, or ejected back into inter-stellar space by the gravitational effects of planets, by jets from the proto-star, or evaporated by ultra-violet radiation from the star when it reaches the T Tauri stage. Once the proto-stellar disk has gone, the exoplanetary system (if any) is more-or less in its final form. There will be a continuing bombardment of the larger planets by smaller objects and this may be very intense for a few 100 million years until the smaller objects are largely mopped-up. There may also be continuing gravitational interactions between the planets that may alter some orbits significantly. Since stars usually form in groups from the condensations within a GMC (like the Trapezium star cluster in the Orion nebula or the Pleiades), then there could be interactions with one or more of those neighbouring stars. A close encounter between two stars is likely to have drastic effects upon any planetary systems. Although direct collisions between planets and/or stars are improbable, the planets' orbits will certainly be changed and some planets could be lost to their parent stars to become free-floating exoplanets. Another possibility is that a nearby star could become a super-nova. If the super nova explodes within the GMC its expanding shock wave could compress some of the gas to the point where new star (and exoplanet) formation is initiated. There is some evidence from the presence of unusual isotopes of some elements within meteorites that the formation of the Sun and solar system was helped in this fashion. At a later stage the radiation from a nearby supernova could sterilise any emerging life-forms that might exist within the fledgling planetary system, significant changes to the planets themselves though are improbable.

Exoplanetary systems that survive these perils are likely still to be around when their host star begins to evolve away from the main sequence. For solar mass stars the main sequence lifetime is around 10,000 million years and it is even longer for lower mass stars. Only some 13,800 million years has elapsed since the universe originated in the Big Bang, so most exoplanetary systems are still orbiting main sequence stars. Nonetheless we may postulate what may happen to planets (including our own) when their host star does begin to change.

What happens to stars after their main sequence lives are over depends upon their masses. The most massive stars usually become supernovae and their exoplanets will mostly be destroyed or blasted out into inter-stellar space during the explosion. Even companion stars have difficulties in surviving a supernova. The current ~10 Jupiter mass object orbiting the milli-second pulsar, XTE J0929-314 was probably once a star with a mass around half that of the Sun (pulsars are the neutron star left-overs of supernovae). Remarkably though, it is possible for exoplanets to survive a supernova as chthonian remnants – the first detected exoplanets, PSR 1257+12 B and PSR 1257+12 C, are probably just such escapees.

No star with a mass less than about 90% that of the Sun has had time to complete its main sequence life yet. However theory suggests that those stars with less than a quarter of the solar mass will simply cool off and become white dwarfs once their internal nuclear reactions cease. Exoplanets of such stars will therefore, rather boringly, stay largely unchanged except for getting colder and colder as their star fades. Eventually their host stars may even become black holes, but this is likely to require a time many hundreds of times longer than the present age of the universe.

Between a quarter and a few times the mass of the Sun, stars (including the Sun) will expand in size and become brighter when they leave the main sequence. Within 10 million years or so they will have become giant stars of various types with the more massive stars becoming super giants and reaching 10,000 times the brightness and 1,000 times the size of our present Sun. Clearly the increased brightness of the host star (even though it will also become cooler and redder) will increase the temperatures of any exoplanets that it possesses. However unless a planet is on the margin between (say) having a solid surface and a liquid one, or retaining and losing an atmosphere, etc., the brightness change of its host star will have relatively little effect. The change in the star's size, though, will drastically affect the inner planets since they will become engulfed by the star. The drag of the outer layers of the star on the planets' orbital motions will lead to them spiralling in towards the star's centre and eventually being destroyed. Since stars' lives as giants are short – typically a few million years – it is possible that the rocky/metallic cores of hot Jupiters and hot super Jupiters could survive even being inside a star and so emerge to continue orbiting the shrinking remnant of the giant once it starts to collapse towards becoming a white dwarf.

A possible such case of a star consuming an exoplanet has been found recently for BP Psc, a probable red giant about a 1,000 light years away from us. The star has been observed by the Chandra spacecraft to be embedded in an extensive disk of gas and dust. A red giant should have long ago lost its proto-stellar nebula so it is suggested that the observed disk is the result of the recent break-up of a large exoplanet or a small companion star arising from interactions with the evolving and expanding primary star. Ironically it is possible that a new set of planets could now be forming within the debris.

In the case of the solar system the Sun is likely to swallow Mercury and Venus during its giant phase, but the Earth may just be far enough out to escape that fate. The Earth though will lose its atmosphere and its surface temperature could rise to 1,000°C or higher – so any surviving descendents of the human race must expect to have to emigrate to the moons of Jupiter or Saturn in around 5,000 million years from now. Exoplanets that escape destruction during the giant phase of their host stars will find themselves orbiting (perhaps in drastically changed orbits) a star that is throwing off a planetary nebula (Figure 13.1b) and collapsing down to a white dwarf. Like the exoplanets of low mass stars they therefore face a future in which they just cool down – eventually, along with their star, reaching the temperature of inter-stellar space.

Complications to the pattern of stellar evolution, and therefore to the futures of the stars' exoplanets, arise in many ways. Close binary star systems are likely to interchange mass as they evolve and may become dwarf, recurrent or classical novae or polars (AM Her stars), etc. Other types of variables such as Miras, R CrB stars, flare stars and so on will also have their effects upon any planets. Even the 'conventional' stars discussed earlier are likely to go through a variable phase becoming Cepheids, W Vir or RR Lyr stars following their giant stages. Exoplanets that manage to survive such stellar convulsions unscathed could still harbour life forms. There may, of course, be extremely hardy ETs able to live through the turmoil in their living conditions resulting from the changes in their host stars. Given, though, that we are currently agonising over whether or not the human race can survive a temperature increase of just a few degrees, it seems more likely that alien life forms will also suffer severely or be rendered extinct on the exoplanets of any but the most stable host stars – but that is a subject for the next chapter.

14. Future Homes for Humankind?

Introduction

Many people see the reason behind the search for exoplanets as lying in the hope the we might find another planet that could provide a second home for ourselves (plus of course providing places for ants, apple trees, butterflies, cats, cattle, cod, dogs, dolphins, elephants, grass, maize, nightingales, oaks, potatoes, redwoods, rice, salmon, sheep, spiders, thrushes, trout, vultures, wheat, etc.).

That second home may be needed soon if we continue to render our first home uninhabitable at the current rate. Even if climate change, pollution and other problems do not destroy the Earth's present-day ecosystem, it would be nice to have a refuge in case of an external threat such as an approaching dinosaur-killer-sized asteroid. Another, often tacit, motive may be the thought that a suitable exoplanet could provide a place for the Earth's increasing human population and provide fresh supplies of resources like oil to replace those that are now rapidly running out.

Exoplanets as safety valves for excess populations or as new resource-providers are a non-starter on numerical grounds, quite apart from any technological, financial, physical, humanitarian or sociological objections that there may be. At most, at enormous cost, we may within the next century or two establish small inhabited outposts on the Moon, Mars and perhaps some of the asteroids or a moon or two of Jupiter (which would at least though, provide some refuges against dinosaur killers). Even if it were to be possible, the emigration of excess population to other planets can only postpone the moment by a few centuries when the population of all planets exceeds the capacity of those planets to support it (Appendix IV). Neither can exoplanets be expected to provide replacements for dwindling terrestrial resources. Even within the solar system the cost in terms of the consumed resources of (say) mining a small asteroid would exceed the value of any useful products by a large factor. The same comment would apply a million million million times over to any attempt to provide any supplies of any material items from even the nearest exoplanet.

Nonetheless, let us examine some of the possibilities of exoplanets as homes or refuges etc., even though at times it will require the use of some little known (i.e. as yet undiscovered) scientific laws and techniques.

Could We (or Un-manned Probes) Ever Travel to an Exoplanet?

!!YES !!
... In fact ... we're already doing just that ...
... But ... (can you wait a few million years?)
We already have spacecraft potentially travelling towards exoplanets. Those spacecraft were not aimed to travel towards specific exoplanets, but are travelling fast enough to depart from the solar system and so progress into interstellar-space where they may encounter exoplanets at some time. The first spacecraft able to leave the solar system are Pioneer 10 and Pioneer 11 launched in 1972 and 1973 to fly-by Jupiter. Pioneer 10 is currently about 100 AU away from the Sun and heading at a speed of about 13 km/s towards a point in the sky about two thirds of the way between α Tau (Aldebaran) and β Tau. The nearest currently known exoplanet in that direction is actually a triple-planet system, HD 37124 b, c and d. The planets are all around 0.6 Jupiter masses and range from 0.5 to 3.2 AU out from their solar-type host star. Pioneer 10 will be in their vicinity about two and a half million years from now. Pioneer 11 flew by both Jupiter and Saturn and is now some 80 AU out heading roughly for α Scu at 11.6 km/s. It will be in the locality of CoRoT-11 b, a 2.3 Jupiter mass hot Jupiter, in 50 million years time. Both Pioneer spacecraft carried plaques (Figure 14.1) intended to let any ETs finding them know who sent them and

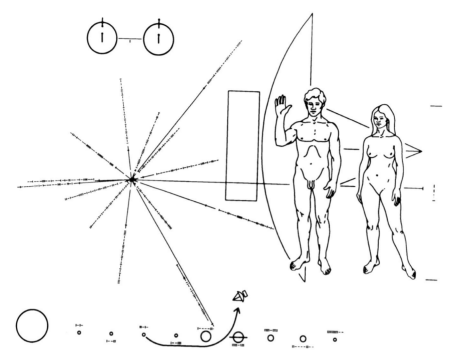

FIGURE 14.1 The 'Information for ETs' plaques carried by Pioneers 10 and 11. (Reproduced by kind permission of NASA).

from where they departed – so as far back as 1972 visits to exoplanets were clearly anticipated.

Three other spacecraft are on their way to inter-stellar space – Voyagers 1 and 2 and the recently launched Pluto mission, New Horizons. The Voyager spacecraft were launched in 1977 to fly-by Jupiter and Saturn and, in the case of Voyager 2, to fly-by Uranus and Neptune as well. Voyager 1 is 114 AU out heading at 17 km/s towards a point about 6° West of α Oph. It will bypass the super Earth exoplanet, GJ 1214 b and its red dwarf host star, in about the year 760,000 AD. Voyager 2, heading in a direction near ν Tel, will similarly bypass the slightly-hotter-than-the-Sun star HD 187085 and its cold Jupiter exoplanet about three million years from now. Like the Pioneers, the Voyagers carry messages for ETs – though this time in the form of gold-plated copper phonograph disks containing sounds and images of the Earth.

After passing Pluto and possibly one or two KBOs New Horizons will head for a region near π Sgr, arriving after two million years in the region of HD 179949 b, a hot Jupiter orbiting a solar-type, but hotter, host star. After another six million years the spacecraft will bypass a couple of exoplanets, HD180902 b and HD 181342 b, which are a cool Jupiter and a cool super-Jupiter that both happen to orbit cool giant stars.

The Pioneers, Voyagers and New Horizons spacecraft demonstrate that our existing technology is sufficient to send missions to exoplanets. These five spacecraft however were never designed (despite the plaques and gold-plated records) for such a purpose and they will cease to operate after traveling only a tiny fraction of the distance to any exoplanet and thereafter just drift through space as derelict hulks. Even if we could manufacture a spacecraft capable of operating for millions of years and then having sufficient power to broadcast information back to the Earth it is inconceivable that such a long-term project would ever be undertaken. Realistically therefore we need to examine whether or not it is possible to send a mission (manned or un-manned) to an exoplanet and obtain results within a few tens of years or maybe a century or two.

Let us start by taking a 100-year un-manned minimalist fly-by mission as an initial goal and call it Project CHEAP (CHickenfeed Expedition to Another Planet). The nearest currently known exoplanet is ε Eri b at a distance of 10.5 light years. Doubtless nearer exoplanets will be found, but 10 light years also seems reasonable as an initial goal. To get results back to the Earth within a 100 years leaves just 90 years travel time. The spacecraft must therefore travel at an average speed of a ninth of the speed of light (33,000 km/s). For a spacecraft massing 1,000 kg, the energy (assuming 100% efficiency) required to give it such a velocity is 500,000,000,000 million joules – a large (1,000 MW) terrestrial power station would take nearly 20 years to generate that amount of energy.

The cost of just the kinetic energy of the CHEAP spacecraft at 2010 domestic rates is around $15,000 million (€12,000 million, £10,000 million). Since no process is ever 100% efficient and there will be other costs as well – design, construction, infra-structure, perhaps the transport of components into Earth orbit, personnel costs etc., a total cost for Project CHEAP of some $100,000 million (€80,000 million, £65,000 million) is probably a minimum estimate.

Large though the cost of the project must be, it is not impossible that it might be funded. The International Space Station is currently costing around $130,000 million (€100,000 million, £80,000 million). Similarly the Apollo moon landings programme cost some $25,400 million in 1974 dollars. Today's equivalent would be around $200,000 million (€160,000 million, £130,000 million). So – if we want to – we could *afford* to send a spacecraft to a nearby exoplanet and get results within a human lifetime.

Even when given a cheque for $100,000 million how do you get a spacecraft up to a speed of 33,000 km/s? In 1903 Konstantin Tsiolkovsky showed that any desired final velocity could be reached by a rocket provided it had enough material to start with and could shoot that material backwards out of the rocket's nozzle at a high enough speed. A chemical rocket such as those in current use has an exhaust velocity of around 5 km/s. Unfortunately Tsiolkovky's work shows that the entire mass of the visible universe would be insufficient at accelerate a single electron to 33,000 km/s if utilized in the form of a simple chemical rocket. Using multi-stage rockets with ion drives (where the exhaust velocity reaches 50 km/s) would still leave us needing a lot more matter than there is in the visible universe. With an exhaust speed almost equal to the speed of light though, the launch mass for a 1,000 kg payload would fall to a more manageable 1,110 kg.

Now we can accelerate protons and other sub-atomic particles to speeds very close to the speed of light in particle accelerators like CERN's Large Hadron Collider (LHC) and Fermi Lab's Tevatron. Unfortunately these accelerators are huge (the LHC is 27 km across) and weigh tens of thousands of tons, so could hardly be included on a 1,000 kg spacecraft.

However light itself (which clearly travels at the speed of light) exerts a force. If you hold your hand up to a beam of sunlight then a force about equal to the weight of a single grain of common table salt will push it away from the Sun. Such a force would be no use to the CHEAP design team, but some of the most powerful lasers in the world are to be found at the Lawrence Livermore National Laboratory in California where they produce pulses of light peaking at intensities of 500,000 MW for tiny fractions of a second in an attempt to start fusion reactions. This is 20,000 million times the amount of sunlight on your hand and exerts a force equal to the weight of six 25 kg bags of salt. Such a laser, if it could operate continuously, would be able to accelerate the Project CHEAP spacecraft to 33,000 km/s in about 12 days. Unfortunately, if not as large as the LHC, the lasers are still huge and with their associated power sources and other equipment again mass many thousands of tons. Luckily, the lasers do not have to be on the spacecraft – they can be on the ground and their light sent to be reflected from a mirror mounted on the spacecraft. Such a system also gives the spacecraft twice the push that it would get from carrying the same lasers on board.

The mirror used to reflect the laser light would have to be huge – at least several kilometres across, perhaps several tens of kilometres across, because even a laser beam that has been focused as precisely as possible will be that wide by the time the spacecraft has travelled as far as Mars' orbit. In fact the mirror is more usually called a sail since the spacecraft is in effect propelled in much the same way as a sailing ship. The reflective sail would be made from extremely thin metal film or metal-coated plastic film and would pull the spacecraft behind it at the end of a long line. The idea of using reflective sails to propel spacecraft utilizing solar radiation rather than light from a laser has been around since the early 1960s and has been the subject of several science fiction stories and films. Arthur C. Clarke wrote of a race between seven solar-sailed spacecraft in 1964 in *The Wind from the Sun*, while 10 years later Larry Niven and Jerry Pournelle envisaged a somewhat scaled-up and manned (actually aliened) version of the Project CHEAP spacecraft in *The Mote in God's Eye*. In the summer of 2010 the non-fictional Japanese spacecraft IKAROS (Interplanetary Kite-craft Accelerated by Radiation Of the Sun) finally demonstrated a working 20-m solar sail which it is now using to propel itself first towards Venus and then to the far side of the Sun. NASA's NanoSail-D spacecraft similarly started operating in Earth orbit in January 2011 with a 10 m² sail.

Thus we could not only afford Project CHEAP, but with currently available technology plus the improvements and developments of that technology that are likely to occur within about the next decade we could also just about manage to accelerate the spacecraft to the speed that is needed. There would then remain a few additional problems such as:

- Ensuring the reliability of the spacecraft over a period of a century
- Developing a suitable power source
- Ensuring that the laser light does not evaporate the reflective sail
- Protecting the spacecraft against impacts (at 33,000 km/s a collision with a dust particle the size of a grain of table salt would be equivalent to 4,000 bullets from a Winchester rifle hitting the spacecraft simultaneously)
- Ensuring that the computers in use when the results of the mission are received can still understand the century-old computers on the spacecraft
- Optimizing the path of the spacecraft through the exoplanetary system (which will have to be a completely automatic process)
- Making the desired measurements and obtaining desired images in the few milli-seconds that the spacecraft will have as it zooms past the exoplanets whilst still moving at 33,000 km/s
- Communicating with the Earth over a distance of ten light years
- And doubtless many many more problems, most of which cannot even be imagined at this stage.

Finding solutions to these undoubtedly very tough problems will not be an easy job but it is one that should prove to be relatively minor compared with the task of getting the spacecraft accelerated to 33,000 km/s. Assuming that we do need another 10 years of development of the required technologies and that (say) 10 years is needed from the go-ahead to the launch of the Project CHEAP spacecraft then it is likely that a few survivors from babies born around the year 2030 will still be living to hear of the results of the first inter-stellar mission to an exoplanet.

Project CHEAP highlights the main difficulty of interstellar travel whether it is towards an exoplanet or any other type of deep space object. The distances to even 'nearby' objects beyond the confines of the solar system are so great that either enormous speeds or enormous times or both are required for the journey. Enormous costs are also involved as the estimates for even the minimal Project CHEAP mission show. Nonetheless considerably more ambitious interstellar missions than that of Project CHEAP have been discussed and designed in some detail.

The first serious attempt to design an inter-stellar spacecraft was Project DAEDALUS in the 1970s. The project involved a dozen or so engineers and scientists led by Alan Bond and was conducted under the auspices of the British Interplanetary Society (http://www.bis-spaceflight.com/). The aim was to send an un-manned probe to Barnard's star (5.94 light years away) using the then existing technology or its near-term foreseeable developments and within a travel time of 50 years. The final design envisaged a two-stage rocket driven by small nuclear fusion explosions. The fuel would be in the form of small pellets containing the heavy isotope of hydrogen (heavy-hydrogen or deuterium) and the light isotope of helium (helium-3). Helium-3 is so rare on Earth that 20 years would be needed before the mission could be launched in order to obtain it from the atmosphere of Jupiter using automated processing plants suspended from hot gas balloons. The deuterium and helium-3 undergo fusion reactions relatively easily and would be fused by beams of electrons, converting about 1% of their mass into energy and driving the remaining 99% of the material out of the reaction chamber at around 10,000 km/s. Two hundred and fifty pellets per second would be detonated. It is perhaps worth noting that even now, over 40 years on from the work on Project DAEDALUS, the basic drive mechanism of fusing deuterium and helium-3 pellets has not been made to work.

The Project DAEDALUS spacecraft would be built in orbit around the Earth and would start off with a mass of 54,000 t (yes – that figure *is* 54,000 t). The payload would be about 500 t, the spacecraft's remaining structure some 3,500 t and the fuel 50,000 t. The first stage rocket would fire for 2.05 years and the second stage rocket for 1.76 years leaving the spacecraft with a final velocity of 36,000 km/s. Like the Project CHEAP spacecraft, the DAEDALUS spacecraft would be a fly-by mission. However up to 7 years before arriving at Barnard's star it would launch a total of 18 smaller spacecraft that would be aimed at individual targets and which would relay their results back to Earth via the main 'mother craft'.

Estimating a cost for Project DAEDALUS is not easy, however a rough attempt can be made. It currently costs around $20,000 to put a 1 kg payload into a low Earth orbit. At the same rate it would cost some $500,000 to put 1 kg into orbit around Jupiter starting from the surface of the Earth. It would cost around $1,700,000 to place 1 kg at Jupiter's cloud tops and about $1,200,000 to transport 1 kg from Jupiter's cloud tops to an orbit around Jupiter. Project DEADALUS would require around 25,000 t of He-3 that had been mined from Jupiter's atmosphere to be carried into orbit around Jupiter. The price for this part of the project alone would be some $30,000,000 million (€25,000,000 million, £20,000,000 million). Add to this the price for the robot factories to mine the He-3 and the cost of transporting them to Jupiter's cloud tops, and the cost of transporting the remainder of the spacecraft to Jupiter (or of moving the He-3 back to the Earth), and the price tag rises to $50,000,000 million. On top of that there will be many other costs – development, personnel, materials, support facilities on the Earth or Moon or in Earth orbit or Jupiter's orbit, etc., and a figure of $200,000,000 million (€160,000,000 million, £130,000,000 million) – 1,000 Apollo programmes or 14 times the entire gross domestic product of the U.S.A. in 2008 – is probably a conservative estimate. Clearly, even if practicable, Project DAEDALUS is *not* affordable.

The British Interplanetary Society together with the Tau Zero Foundation (http://www.tauzero.aero/) is currently working on Project ICARUS which is to be an up-date of Project DAEDALUS and which has the aim of a potential launch date before the end of this century. It is proposed to send a spacecraft to an exoplanetary system within 15 light years of the Earth and for the payload to decelerate at the star. The aim is to have a mission duration of less than a century. So far the only exoplanetary systems detected within 15 light years are ε Eri (10.5 light years) which possesses a cold Jupiter, a circumstellar disk and possibly a second, more distant, exoplanet plus possible asteroid belts and GJ 674 (14.8 light years) that has a smallish warm or hot Neptune.

Another concept which had much in common with Project DAEDALUS was NASA and the US Naval Academy's Project LONGSHOT. This was for an un-manned mission to α Centauri in which the spacecraft would slow down and go into orbit around the star. The propulsion method was again to be mini-fusion explosions of deuterium and helium-3 pellets although a separate fission reactor would provide the power for running the spacecraft. The initial mass would be 400 t and the payload 30 t and it would have a flight time of 100 years at 13,000 km/s.

A rather more hair-raising concept for an inter-stellar space-craft's propulsion system that could be made today is the use of 'proper' nuclear explosions. In the 1950s and 1960s Project Orion envisaged a spacecraft mounted via shock absorbers onto a massive steel plate. Small nuclear bombs would be exploded behind the plate and the expanding plasma clouds would push the space-craft to ever higher velocities. The design originally envisaged a 4,000 t spacecraft being launched from the surface of the Earth with the nuclear explosions equivalent to 150 t of TNT. Each explosion would add about 50 km/h to the vessel's speed so that 800 explosions would be needed to reach low Earth-orbit. However such a craft would then have been able to undertake a return journey to Pluto in less than a year. In 1968 Freeman Dyson calculated that an Orion-type spacecraft using one megaton explosions could reach α Centauri in 1000 years. It would have a launch mass of 40,000,000 t, carry 30,000,000 bombs, have a payload of 5,000,000 t and reach a speed of 1,000 km/s. The related Medusa concept produced by the British Interplanetary Society in the 1990s also had much in common with the solar sail propulsion method. A large parachute would replace the pusher plate of the Orion spacecraft and the payload would be carried at the end of a long tether. The nuclear explosions would be set off between the payload and the parachute with the parachute catching much of the expanding gas cloud from the explosions and pulling the payload along behind it.

Some readers may now be feeling that they have mistakenly picked up a science fiction book. However Projects CHEAP, DAEDALUS, LONGSHOT, ORION and MEDUSA are all just-about possible or ought to be within a few decades. Whether or not they are affordable or desirable is a quite different consideration. These projects however do represent the limits of what present day science has to say about the possibilities of inter-stellar travel. To explore the possibilities any further we have genuinely to enter the realms of science fiction. Of course many science fiction writers by-pass the difficulty of travelling vast distances by inventing faster-than-light spacecraft drives or hyper-space or warp factors or teleportation or maximum improbability drives or time travel (which can convert any propulsion system to faster than light by traveling back in time throughout the journey) or 'beam me up Scotty' or .... etc. However some science fiction writers do develop ideas that have some realistic potential to them.

One of the most attractive of such ideas that has been used in many SF stories was first proposed by Robert Bussard, an American nuclear physicist, in 1960. He suggested that because most of the hydrogen atoms in inter-stellar space are ionised (i.e. have lost their single electrons and exist as the electrically positive sub-atomic particles called protons) a suitable magnetic or electric field could gather them up. Once collected the protons could be fused together to form helium nuclei and so release large amounts of energy (carbon nuclei would probably be needed as catalysts in this reaction). The energy could then be used to accelerate the helium nuclei to high velocities and so produce a rocket. The device is now known as the Bussard Ramjet and it has the huge advantage of picking up its fuel along its journey – the vast launch masses needed for the DAEDALUS and ORION spacecraft are thus no longer necessary. Furthermore a working Bussard ramjet would allow continuous acceleration on the outward half of the journey followed by continuous deceleration over the second half. Even if the acceleration is only 10% of that of the Earth's gravity (0.1 g) after 5 years the spacecraft would be travelling at half the speed of light (relativistic effects would probably prevent much higher velocities being reached). Project CHEAP's ten light year journey would thus take 25 years instead of 90. Unfortunately the Bussard ramjet has two problems. The first is that we are decades, maybe centuries, away from being able to fuse enough protons into helium nuclei to provide significant amounts of energy. The second is that the collection of the protons by the magnetic or electric scoop will act as a drag on the spacecraft – at least slowing it down and perhaps, if the drag exceeds the thrust, stopping it from working at all.

Anti-matter is also widely used by SF writers as a power source. All sub-atomic particles have 'evil twins' that are reversed in all their properties (electrically positive instead of negative etc.). The positron is the anti-particle of the electron, the anti-proton the anti-particle of the proton and so on. Anti particles can be produced in nuclear reactors and particle accelerators and also result from some types of radioactive decay. When a particle and its anti-particle meet up they annihilate each other completely and produce a pair of gamma rays. By putting together an anti-proton and a positron physicists have managed to make an atom of anti-hydrogen. Anti-matter could therefore in theory be produced simply by manufacturing many billions of anti-hydrogen atoms (or anti-oxygen atoms, or anti-iron atoms, etc.). The anti-matter would have to be strictly isolated from ordinary matter by being suspended using magnetic or electric fields within a vacuum chamber. However when energy is needed a small amount of anti-matter can be mixed with a similar amount of ordinary matter and both types of matter will be completely converted into energy. Thus a kilogram of matter and antimatter would convert into 90,000,000,000 million joules and just 6 kg of the mix would be sufficient to provide all the energy needed by the Project CHEAP spacecraft. Matter – anti-matter annihilation is the most efficient way possible to convert matter into energy since the conversion rate is 100%. For comparison hydrogen fusion has an efficiency of about 0.6%, the fissioning of uranium has an efficiency of 0.1% while the burning of coal or oil or rocket fuel has an efficiency in the region of 0.000,000,1%. Provided therefore that safe storage of antimatter can be arranged it would provide a compact and lightweight source of energy that could be used to power other propulsion systems (rockets, launching lasers, nuclear bombs etc.). The drawback of antimatter is that at the moment it takes billions of times more energy to create the antimatter than would be generated by mixing the antimatter with normal matter.

What of faster-than-light travel? Everyone 'knows' that Einstein's theories of relativity say that it isn't possible. However that 'knowledge' is not quite right – it is actually traveling *at* the speed of light that is forbidden – faster-than-light movement is allowed. In fact physicists have speculated that there may be particles, called tachyons, which always travel faster than light, although no sign of their existence has yet been detected.

In order to travel faster than light then, perhaps all that one has to do is to accelerate to beyond the speed of light *without* ever attaining the speed of light itself. This seemingly impossible task may be made possible with the help of quantum mechanics. Many sub-atomic processes including radioactivity are known wherein a particle can jump from one place to another without passing through the intervening space. This phenomenon arises because under quantum mechanics the particle has some probability of being in both places. So perhaps there might be a quantum mechanical jump from sub-light to super-light speeds that would do the trick.

Then again Einstein's theories *are* theories – and just as Einstein's equations replaced those of Newton and Newton's replaced those of Kepler, so, some day, there will be a replacement for the theories of relativity – and perhaps the replacement will show us how to exceed the speed of light. Whether or not tachyons exist and, if they do exist, whether or not we could somehow make use of them to travel faster-than-light seems likely however to remain unknown for a great many years yet, so that we are currently stuck with the sub-light methods of getting to the stars.

One problem with any project designed to take a century or thereabouts to travel to a nearby star or exoplanet is that during the time that it takes the spacecraft to accomplish the journey new discoveries and inventions may have been found that enable the journey to be made in half (or a quarter or a tenth or a 100th or ...) of the time. Alternatively and perhaps more probably, while the discoveries and inventions may not speed up the journey, they may reduce the cost by similar large factors. There will thus always be a plausible argument for putting off an interstellar mission until it becomes quicker or cheaper or both.

A related argument may be summarized as 'why go to another star anyway?'. The main reason for un-manned interstellar travel, at least, is to obtain information about the star or planet or planetary system etc. It is more than possible, given the sums that need to be expended for an interstellar journey, that that same information could be obtained more quickly and more cheaply by developing and improving existing remote sensing instruments on the Earth or within the solar system. For example some of the missions proposed in Chap. 10 may seem fantastic but their costs are trivial compared with Project DAEDALUS etc. The New Worlds Imager with a fleet of (say) ten spacecraft at $1,000 million each could provide direct images of terrestrial-sized exoplanets for a tenth of the cost of Project CHEAP. It could be started now with existing technology, it would give us results within a decade *and* it could look at numerous exoplanets, not just one or two.

The prospects for an un-manned mission to an exoplanet, for a variety of reasons, thus seem very poor. Surely therefore the prospects for a manned mission must be far worse? Well – Yes and No. The costs of such a mission are probably comparable with those of Project DAEDALUS, but the motivation would be different and while there would be many new problems, some of the older ones might become less significant.

Any manned mission to an exoplanet (barring the invention of hyperspace travel, etc.) would be a one-way journey with the primary aim of establishing a viable colony of human beings on the exoplanet. As already discussed (see also Appendix IV) the purpose of a colony on an exoplanet would not be to reduce population pressure on the Earth, but to provide a back-up for the survival of the human race should it become extinct within the solar system. Evolution has spent 4,000 million years developing the reproductive urge in terrestrial living organisms so the prospect of being able to ensure a greater hope for the survival of one's genes into the distant future could provide an adequate motive for making the effort required for an inter-stellar expedition.

The principal problem with un-manned missions is the requirement for enormous speeds if results are to be returned within a human lifetime or thereabouts. A mission to colonize an exoplanet might not have the same urgency and so could take much longer over the journey. Undoubtedly the 'folks back home' would like to hear about the success or otherwise of a colonizing mission quickly, but the purpose of the mission would still be fulfilled even if it took a 1,000, 10,000 or a 100,000 years over the job.

The main possible methods of sending out a colony to an exoplanet have been extensively discussed by science fiction authors and divide into Arks, Hibernations, Embryo Transfer and Data Exchange. The last of these requires a friendly ET at the other end who has the ability to build up a human being from atoms and molecules, or at least to build up our genes and then provide for more normal reproduction, based upon information broadcast to him / her / it from Earth. Quite apart from providing a hostage to fortune (suppose the ET was a gourmet with an appetite for human flesh?) it does not seem likely that it would be a way of establishing a colony. Let us therefore look at the possibilities of arks, hibernations and embryo transfer. The exoplanets in these cases would have to be twin-Earths if the colony were to be established without modifying the planet in some way (see terra-forming below) and so journeys of hundreds of light years are likely to be needed.

An interstellar ark, or generation ship, would take several, perhaps tens or hundreds, of human generations over its journey, travelling at a speed of 1,000 km/s or less. It would have to be large enough to contain some thousands of inhabitants in an environment that provided for all their needs over at least centuries and probably over millennia. The inhabitants would be born, grow-up, be educated, reproduce, work and die on board the ship until it reached its destination when they would (somehow) transfer to the designated exoplanet. Given the rate of change in human societies on the Earth, there is clearly no guarantee that the survivors would be able to or perhaps to wish to leave their cosy home for some wild exoplanet at the end of their journey – in *Orphans of the Sky*, for example, Robert Heinlein speculates about such a generation ship aimed at Proxima Centauri wherein the inhabitants have completely forgotten that they are even on board a ship by the time that they arrive at their destination. There is also a major ethical question, especially for those inhabitants who are born, live and die during the journey without experiencing either the Earth or the destination exoplanet, that they are the non-consenting agents (slaves might be a better word) of the project's originators with little in the way of free will or choice.

Since a generation ship has much in common with an entire planet in the way that it provides a habitat, some science fiction authors, notably Larry Niven in the *Ringworld* series and James Blish in the *Cities in flight* series have resorted to moving entire planets as a means of traveling around the universe. A more ethical variation on this theme would be possible if human lifetimes were somehow to be greatly extended. It might then be possible for the original travelers to survive to the end of their journey. The prospects though of extending human lifetimes to the thousands of years that would be needed seem to be extremely remote.

A hibernation, frozen sleep or suspended animation trip to another star would have the merit of avoiding the ethical question posed for the inhabitants of arks, since the original, presumably consenting, travelers would be the ones to arrive at their destination. There is little to add to this approach save that some method of causing human beings to go into a state of long-term hibernation needs to be discovered before it can be of any use. Two possibilities for inducing hibernation currently seem to be available – a reduction in body temperature, perhaps including replacement of the blood with another fluid or the use of chemicals such as hydrogen sulphide to reduce the heart beat. Both of these approaches can have side effects such as brain damage and only slow down, but do not halt, the life processes. However further research may produce a system whereby a group of people could be placed in suspended animation for a few hundred years, sent via a slow spacecraft to an exoplanet, to be automatically revived upon reaching their destination and then going about their mission.

The third approach – embryo transfer – is the only one that might be possible within a century or two. Embryos or eggs and sperm are already preserved for years by freezing them in liquid nitrogen. 'All' that is therefore still required before a viable inter-stellar colonizing expedition becomes possible is to devise a robotic system for incubating those embryos and then rearing the resulting babies to adulthood on arrival at their exoplanet. There is still the ethical question of the embryos' lack of consent to this process but it is perhaps a less pointed one than was the case on board the inter-stellar arks since no embryo currently has a choice about its destination and its situation after its birth. There is also a clearly likely to be a major social problem in how the first generation will be brought up in the absence of parents and older exemplars, but doubtless this can be solved through the use of sophisticated robots and the like. Given then a sufficiently reliable supply of liquid nitrogen and sufficiently sophisticated computers, a spacecraft carrying thousands of human and other species' embryos could be launched at a speed little more than that of Pioneers I and II to arrive at an exoplanet in a few million years time and at a cost perhaps a few times $10,000 million.

Whether these rather fanciful speculations have any chance of being realized in some form in the future is cast into some doubt by the Fermi paradox. In 1950 Enrico Fermi asked why we see no sign of ETs inhabiting suitable spots throughout the galaxy. He pointed out that even at the relatively slow speed of 300 km/s, it would take an alien civilization just 100 million years to expand throughout the Milky Way galaxy. The galaxy is at least a 100 times older than this so that intelligent beings that evolved early in the life of the galaxy would have had plenty of time to occupy all the best spots long before we arrived. There are a number of possible answers to Fermi's question and most of them are pretty depressing.

One answer suggests that the requirement for life to start are so strict and difficult to realise that *we* are the first intelligent life forms to evolve in the galaxy, another that intelligent life self-destructs through over use of resources in a short time, yet another possibility is that aliens are indeed out there, but that they are hiding or that we are not looking for them in the right way. However it could be, given what we have just seen of the colossal difficulties and incredible costs of inter-stellar journeys, that no one has yet thought it worthwhile to make the effort.

A Beginner's Guide to Gardening on Mars: or Could We *Really* Live on Another Planet?

In discussing the prospects of establishing a colony on a planet beyond the solar system we made the assumption that the chosen planet was a twin-Earth. That is, it should be a planet entirely suitable for immediate occupation by human beings without needing any significant modifications. Of course a twin-Earth might well already be occupied by ETs, but that's another matter. Adapting a not-quite-suitable planet (terra-forming – see below) to be suitable for human occupation may or may not be possible but it is certain that it would take a considerable amount of time. After 1,000 year (or whatever) trip through space the prospective colonists are unlikely to want to hang around for a few more centuries whilst their planet is made ready for them. We may take it therefore that *if* an interstellar colonization expedition is ever sent off, it will be to an exoplanet that *can* be lived on.

Closer to home we have a selection of places within the solar system that we might consider for occupation. Places such as our Moon, Mars, some asteroids and some of the satellites of Jupiter and Saturn would certainly be inhabited before interstellar colonization would ever be attempted. Clearly the Moon can be lived upon – Neil Armstrong and 11 other astronauts have done so. However temporary visitations such those made during the Apollo programme are very different from a self-contained and self-supporting colony. Such a colony would need to be able to supply itself with oxygen, water and food as the barest minimum. While water in small quantities is present on the Moon and in large quantities on Mars and some planetary satellites, oxygen and food are not immediately available. However given an energy supply, water molecules can be split to provide oxygen and hydroponics used to produce food. Even if water is not available, oxygen can be extracted from the oxides present in rocks and combined with hydrogen obtained from methane, ammonia or even directly from the solar wind in order to supply it. Thus the true basic require-ment for a colony almost anywhere is an abundant supply of cheap energy.

Once we go beyond the Earth, energy sources are restricted to fission, fusion and solar power. Fusion is not yet viable as an energy source even on the Earth and fission requires a major indus-trial complex to extract and concentrate the uranium-235 that is needed. Thus solar power is the likely energy source for newly established colonies in the foreseeable future. Currently and under experimental conditions solar power cells can convert sunlight into electricity with an efficiency of about 40%. It is probably rea-sonable to expect 50% conversion efficiency in cheap off-the-shelf solar power cells by the time we are ready to establish colonies beyond the Earth. At the distance of the Earth from the Sun and without an atmosphere, sunlight provides 1.4 kW/m^3, near Mars this falls to 0.6 kW/m^3 and at Jupiter's distance from the Sun to 50 W/m^2. Thus a solar panel array to supply a megawatt of power would be about 40 m^2 for a lunar colony, 60 m^2 for a Martian colony and 200 m^2 for colonies on the Jovian satellites. Since many types of solar power cells are made from silicon and this is abundantly available from rocks, producing the cells should be possible on site given a small start-up manufacturing plant supplied from the Earth (recently solar cells have been produced by printing them directly onto a paper substrate). Current costs for domestic solar power cells are around $10,000 (€8,000, £6,000) per installed kilowatt, but we may anticipate this price falling very considerably with increased conversion efficiency and with large-scale production. At $1,000 per square metre (say) the price of a megawatt solar power supply would be $1.6 million for the lunar colony, $3.6 million for a Martian colony and $40 million for the Jovian satellites. Compared with the price tag ($100,000 million) for Project CHEAP these are petty cash sums.

Once the energy supply for a colony has been established then producing oxygen is straightforward. In 1 h, at a 50% efficiency rate, one megawatt will produce sufficient oxygen by splitting water to supply the daily needs for over 100 people. With water and oxygen provided, the colony can turn its attention to food production. For colonies on the Moon or Mars crops may be grown relatively normally using sunlight (probably with some of the ultra-violet light filtered out) under the sort of domes beloved of 1930s SF illustrators. The domes would be plastic bubbles supported by the pressure of the gas inside them – and micrometeorite holes would just have sticky patches applied over them whenever they occurred. Probably crops would be grown hydroponically, but soil could be manufactured by finely grinding up rocks if needed. As an example of what is possible, the McMurdo station in Antarctica currently runs a 60m^2 hydroponic greenhouse producing lettuce, spinach, tomatoes and cucumbers. For Mars, at least, some terrestrial forms of life might be able to survive even without such protection (or genetic engineering could adapt them to do so). Given a bit of fertilizer, a cotton grass found on the arctic island of Spitsbergen for example, could probably grow on some parts of Mars today. The bacteria *Deinococcus radiodurans* and a strain of *Brevundimonas* found in Antarctica have been shown in terrestrial simulations of Martian surface conditions to be likely to be able to survive on the planet in dormant forms for over a million and for a hundred thousand years respectively.

Some form of heating would be required for the lunar nights and for the Martian winters but simple heat storage systems would probably suffice for this. The colonists would probably build their living quarters underground (as anticipated in many SF stories) where temperature variations would be minimized and protection provided from meteorites, ultra-violet radiation, cosmic rays, solar storms and so on. On an asteroid or Jovian or Saturnian satellite the intensity of sunlight would be too low to grow crops directly so that artificial illumination would be needed, otherwise little would change from a Lunar or Martian colony. With oxygen, water *and* food provided the colonists could turn their attentions to lower priorities such as building factories, establishing educational and medical facilities, exploring, expanding the colony, etc.

Terra-forming

Colonies such as those just imagined are a poor way of living away from the Earth – they require humans to adapt to the conditions provided by the planet. Much better to adapt the planet to provide the conditions desired by humans – and that is what terra-forming aims to do.

We are right now demonstrating that terra-forming is a possibility by the climate changes that human activities are causing to the Earth. If we had an exoplanet that was similar to the Earth but just a few degrees too cold, we would know that we could terra-form it to a more comfortable temperature by adding carbon dioxide and methane to its atmosphere so strengthening the greenhouse effect (Appendix IV). If the exoplanet had too much ozone in its atmosphere than adding quite small quantities of chlorofluorocarbon compounds (CFCs) would soon sort out the problem.

As we have seen earlier it seems most improbable that an interstellar colony would be sent to a planet that required terraforming before it could be inhabited. Thus in practice, if terraforming is ever to be undertaken, it will be within the solar system. In the 1960s Carl Sagan and others reasonably seriously proposed that Venus could be made more habitable by employing genetically engineered bacteria or algae to reduce the amount of carbon dioxide in Venus' atmosphere by converting it to organic compounds. Thus a few spacecraft loads of such bacteria dropped into Venus' cloud tops and left to get on with it would be all that was needed. Unfortunately improved knowledge of Venus – in particular the enormous depth of the atmosphere, its very high temperature, the scarcity of hydrogen to form organic compounds and the presence of sulphuric acid in the clouds – shows such a simple approach to be unworkable.

Venus though is similar to the Earth in size and mass and so would make an attractive choice if some other way of terraforming it could be devised. The main first step would have to be to reduce the present atmosphere – currently over 90 times thicker than that of the Earth – to something comparable with the Earth's atmosphere. The quantity of material to be removed in some way in order to reduce the atmosphere to such an extent is around 400,000,000,000 million tons. If bacteria or algae are ruled out and human-devised factories that took in carbon dioxide and converted it to (say) limestone seem equally unlikely, then perhaps the best hope would be to blast the atmosphere into space. Altering the orbit of one of the larger asteroids (Ceres, Pallas or Vesta) so that it collided with Venus would provide sufficient energy in the collision to do the job although ensuring that the energy went into removing the atmosphere and not into producing an enormous crater might be a problem. However as numerous SF books and movies have shown, deflecting even a very small asteroid or comet head from a collision course with the Earth is perhaps about the upper limit of our capabilities at the moment in the planet moving business.

Another approach might be to lower the temperature of the atmosphere by reflecting sunlight away from the planet until bacteria or algae can survive. 60,000 million hundred-metre diameter balloons with aluminized reflective coatings on their envelopes and floating high in Venus' atmosphere might be sufficient for this. Alternatively a single mirror in space could reflect the sunlight away before it even reached the planet. However that mirror, although it could be made of a very thin reflective film like a solar sail (see Project CHEAP), would have to be as large as Venus, if not larger. The forces acting upon a mirror of that size from solar light pressure and the solar wind would be several hundred thousand tonnes, so controlling it and keeping it in position would be impossible. All things considered, trying to terraform Venus seems likely to be at least as expensive as Project DAEDALUS and would probably fail even if that amount of money were to be found.

A few years ago Mars was also thought to be simple to terraform – just heat it up so that its carbon dioxide-containing polar caps evaporate and that gas would then raise the temperature further via its greenhouse effect. The initial raising of the temperature might be effected by the use of large mirrors in space to reflect sunlight onto the planet. Although this is the opposite use of space mirrors from that proposed for Venus, the same comments about the difficulties involved in controlling them would apply. Alternatively gases such as octafluoropropane would induce a runaway greenhouse effect that might melt the polar caps if added to the Martian atmosphere at 300 parts per million. Unfortunately the amount of carbon dioxide locked into Mars polar caps is now known to be insufficient to raise the planet's surface temperature by more than a few degrees.

A variant upon Carl Sagan's algae might though be a more viable possibility. Water is present in the form of ice in plentiful quantities at many places on Mars so it may be possible, via genetic engineering, to produce bacteria or algae etc. able to survive the present Martian conditions. Those organisms by producing more carbon dioxide (or other greenhouse gases such as methane or ammonia) might then over the following centuries or millennia add sufficiently to the greenhouse effect to warm Mars surface to livable temperatures. Such a project might be aided by adding water and/or ammonia from an asteroid rich in those substances. Although a much smaller asteroid would be needed than that suggested for removing Venus' atmosphere, colliding a 10 km diameter asteroid containing (say) 10% of its mass in the form of ammonia with Mars would only increase the present atmospheric pressure of Mars by about 2% – and moving a 10-km diameter asteroid is still well beyond our foreseeable capabilities.

A slightly more viable variant on the asteroid collision idea might be to direct the impact in such a way that it provokes some of the gigantic Martian volcanoes back into activity. Terrestrial volcanoes emit large quantities of gases, so a smallish asteroid collision might suffice to bring the Martian atmospheric pressure up to that of the Earth from volcanic emissions (a pressure increase by a factor of nearly 200). Mars' inner satellite, Phobos might make a suitable projectile, though at over 20 km across it is a bit large. Phobos orbits only 6,000 km above Mar's surface and its orbit is already reducing in size at a rate that in the natural course of events might lead it to crash into Mars in about 50 million years from now. Thus aspiring terraformers might just need to encourage the process a little so that it happens sooner.

Unfortunately any or all of these suggestions, even if successful, result in a Martian atmosphere of carbon dioxide and/or methane and/or ammonia and/or sulphur dioxide etc. So another stage of terraforming, perhaps with yet more genetically engineered bacteria, will be required to produce the oxygen needed for a breathable atmosphere. Terraforming Mars might be just a little more feasible than terraforming Venus – but don't expect it to happen soon.

Of the remaining solar system objects which might be candidates for terraforming, only Titan currently retains a significant atmosphere. That atmosphere is largely nitrogen and methane. Warming it up from its present temperature of –190°C might be possible with space mirrors, although at nearly 10 AU out from the Sun the mirrors would need to be ten times larger than Titan's 5,150 km diameter (i.e. a 100 times its area) if terrestrial temperatures are to be achieved. Given though that the slightly larger and warmer (5,262 km, –100°C) Ganymede does not have an atmosphere, warming up Titan might simply cause it to lose its atmosphere. If warming Titan does not lead to loss of its atmosphere – or what would be almost as good, if that loss takes millions of years to occur – then Sagan's genetically engineered bacteria might finally come into their own and produce the oxygen needed to make the warmed-up surface of Titan a genuine shirt-sleeve environment for human beings.

Real terraforming in the sense that might be possible for Titan and perhaps Mars is not likely to found elsewhere because the objects involved would be of too low a mass to retain an atmosphere at terrestrial temperatures. A partial terraforming might still be possible though in which a thin atmosphere that does not necessarily contain oxygen and which might have to be replenished at regular intervals is generated to help trap some heat and to provide protection against micro meteorites, ultra-violet radiation and particles from the Sun. Although space suits would still be needed outside any enclosed accommodation on such a body, they might amount to little more than the wet suits and scuba gear currently worn by terrestrial divers. The larger asteroids and the satellites of Jupiter might at some future date be candidates for such partial terraforming. The Cassini spacecraft has recently detected the presence of a very tenuous atmosphere for Saturn's satellite, Rhea. The atmosphere contains free oxygen, probably produced by charged particles interacting with water ice on Rhea's surface. However the amount of oxygen is around 0.000,000,000,5% of that present in the Earth's atmosphere, so is of little help in rendering Rhea potentially habitable by human beings.

No Vacancies: The ETs Got There First

Suppose against all the odds discussed previously we finally found a nice exoplanet and sent an inter-stellar expedition to it, but when we got there found that it was already fully occupied by ETs (aka Aliens, Little Green Men (LGMs), etc.). Would the ETs welcome the visitors, ignore them or do their best to get rid of them? Judging by the reactions attributed to human beings by SF writers when writing of the reverse situation (i.e. an invasion of Earth by ETs) the third reaction is the one we should expect. So what might those ETs be like and how probable is it that we might encounter them?

The latter question is perhaps the easier to answer since active programmes have been underway on the Earth to try and detect signals from ETs (SETI – Search for Extra-Terrestrial Intelligence) since the 1960s and earlier. Since even a cursory review of SETI would result in a book at least as long as this one, the interested reader is therefore referred to specialized sources (including the internet – see http://www.seti.org/ and http://en.wikipedia.org/wiki/SETI – for example) for further information. For the purposes of this book it suffices to note that despite very considerable efforts over several decades no signal attributable in any way to an ET of any sort has been received by human beings. One calculation due to Frank Drake (the Drake equation) suggests that at any given moment that might be as few as two or three ET civilizations with whom we might be able to make contact throughout the entire Milky Way galaxy. The implications of this are either that extra-terrestrial intelligences are rare, or that they are there but we have not found them yet or that we are not looking for them in the right way – or a combination of these possibilities.

We only have the development of life on one planet – the Earth – as an example of what is possible and extrapolation from a single case is always likely to be highly misleading. Nonetheless we have to make the most of what we have got. In the case of the Earth, life in some form or other has been around for at least 3,500 million years, life-forms with intelligences comparable with a modern dog have been around for several 100 million years and life forms with intelligences approaching our own have been around for several million years. The species of Homo Sapiens itself has been present upon the Earth for some hundreds of thousands of years.

For an ET searching the Earth or the solar system from a distance of several light years for signs of life here, the first indication would probably have been the increase in the proportion of free oxygen (produced by plant life) in the atmosphere and that commenced around 2,000 million years ago. It is possible that other signs such as colour changes or the spectrum of chlorophyll might also have been detectable at about the same time given observational capabilities for the ET comparable with those of our own today. The first external signs of *intelligent* life (and indeed of any form of animal life) would probably have been the heat and light emissions from the first organized communities, such as Babylon and Petra, some 4–5,000 years ago, although these would not have been detectable at several light years distance. The ET might have been able to detect some of the gases produced by early industries in spectra of the Earth's atmosphere from the eighteenth century onwards. These indications though are all difficult to observe and could arise from inanimate processes.

Thus the first definitive sign of intelligent life on the Earth would have been the development of artificially produced radio waves and this would date from Heinrich Hertz' work in 1888. For longer than the past century therefore to any ET with a radio or microwave telescope the Earth will have stood out like a beacon signaling life's presence here. It seems unlikely that radio and microwave use will disappear in the future, but the usage may decrease as broadcast networks are replaced by cable and fibre-optic links – a development that we are already seeing with the internet. It may therefore be that the radio beacon that is the Earth at the moment will gradually dim over the next few decades.

Thus to the ET observers of the Earth, the presence of life upon it may have been marginally detectable for about 50% of the planet's 4,500 million year life span, but the signs of intelligent life have only been clear for 0.000,003% of the time and may disappear shortly. If (and it's a 'big if') the same scenario applies to life developing on exoplanets, then taking a sample of 20 million exoplanets that *do* have life on them only *one* would currently be detectable by ourselves. Of course, scientific developments by ETs may take different routes from our own so that instead of using radio they are communicating via neutrinos or gravity waves or sonar (for ocean-based life forms) or via some as yet completely unknown process – in which case we have no chance at all of detecting their presence.

Our chances of encountering life on exoplanets (whether of the intelligent variety or not) will also depend upon the chances of life developing within a suitable environment. Estimates of this range from that it has happened just once throughout the entire universe, to that it follows inevitably and swiftly anywhere and everywhere as soon as conditions are appropriate. A couple of lines of evidence suggest that the latter expectation might be the closer to the truth. Firstly, amino acids, which are the basic building blocks of terrestrial life, can be produced inorganically by many processes – they can even be found in inter-stellar space. Secondly a meteorite from Mars, found on the Earth, contains the mineral magnetite in a form that on the Earth is only produced through the action of a certain type of bacterium, thus suggesting that it might have been produced long ago by Martian bacteria. For the moment though it is anyone's guess where the reality lies.

Speculation about other types of life based upon silicon, say, rather than carbon, still lies within the realm of science fiction and is certainly beyond the remit of this book.

Even if life does develop easily, the right conditions for it probably occur rarely. Despite certain organisms able to survive in Antarctica's ice or in volcanic pools of boiling water, most terrestrial-type life depends upon the availability of a supply of liquid water at moderate temperatures. Erring on the generous side, this implies that the temperature of the site where life is to be initiated should lie between around –10°C and 90°C. It may be possible to find parts of some planets that fulfill these conditions – say a few thousand kilometres below Jupiter's cloud tops for example – even though the planet is generally inimical to life.

If a large part of a planet is to be suitable for life it must lie at an appropriate distance from its host star so that its temperature generally falls within the required range. Since the temperature of the surface of the planet will also depend upon whether or not it has an atmosphere and if it does how strong the resulting greenhouse effect may be, whether or not there are clouds present in the atmosphere to reflect the host star's energy away from the planet, possibly upon whether there are other heat sources such as volcanism and even upon whether or not there are deep oceans to act as heat reservoirs, there are a range of suitable distances from the host stars for potentially habitable exoplanets. The range of distances is called the habitable zone or Goldolocks zone (by analogy with the temperature of Baby Bear's porridge – which was 'just right' – in the tale of 'Goldilocks and the Three Bears') of the host star. For the Sun the Goldolocks zone extends from near Venus to Mars or beyond. Clearly for faint, cool, red dwarfs their Goldolocks zones will be narrower and much closer to the stars, while those of hot bright stars will be broader and further out.

Despite some 500 exoplanets now being known, very few are planets within or even close to Goldilocks zones. The confirmed exoplanets, Gliese 581 c and Gliese 581 d (Figure 1.1), lie on the edges of Gliese 581's habitable zone, but have masses five and a half and seven times that of the Earth. The as yet unconfirmed exoplanet in the same system, Gliese 581 g, however lies in the middle of the habitable zone and may have a mass as low as 3 times that of the Earth. If Gliese 581 g is confirmed and if it possesses the other requirements for life then the existence of some form of life upon it must be considered a distinct possibility. Similarly, another possible hopeful is the very recently announced, but also unconfirmed, Goldolocks zone exoplanet candidate KOI 326.01. Its mass may be as low as 80% of that of the Earth but it is likely to have a surface temperature (if, indeed, it has a solid surface) in excess of 60°C.

Amongst the 1,600 exoplanetary candidates detected by the Kepler mission, 44 are within or close to their stars' habitable zones. The Kepler team expect that eventually around 80% of exoplanetary candidates will be confirmed to be genuine exoplanets, suggesting that in due course perhaps 4–5% of exoplanets will be found to lie within habitable zones.

So, to return to the second question posed at the beginning of this section 'how probable is it that we might encounter alien life?'

- If by 'encounter' a direct physical meeting between ourselves and the alien life forms is intended then the probability is effectively zero unless indigenous life or the fossilized remnants of it should be found on Mars or elsewhere within the solar system
- If we take 'encounter' to mean 'detect any form of life's presence on an exoplanet' then the probability is high – it would be disappointing if such a detection has not been made by the time the ten thousandth exoplanet is found (sometime around 2020–2045 – see Figure 3.18
- If by 'encounter' the detection of intelligent extraterrestrial life is meant, then that could be a long way into the future but will probably happen eventually.

The first question in this section 'what might ETs be like?' has been answered hundreds of times in SF books and movies – with the answers being wrong in every single case! The variety of life upon the Earth suggests that alien life forms will be an even more disparate assemblage, however evolution will always tend to make life forms suited to their environment and way of life. On the Earth we have sharks (cartilaginous fish), tuna (bony fish) and dolphins (mammals) that have evolved quite separately to live in the oceans hunting other animals for food. As a result of the requirement to move rapidly through water, all three groups of animals have evolved similar body shapes. Alien life that lives and hunts in alien oceans (whether these be formed from water, methane, ammonia, or some other liquid) would thus seem quite likely to develop a similar body shape to these terrestrial examples. Similarly bats and birds strongly resemble each other, with small lightweight bodies and paired wings – because they both have life styles involving flight through a gas. Yet again marsupial and non-marsupial animals have evolved separately from each other for many millions of years but similar lifestyles led to marsupial lions (Thylacoleo carnifex) and the Thylacine (marsupial wolf or hyena – Thylacinus cynocephalus) that strongly matched 'conventional' lions, tigers, wolves and hyenas in their appearance.

Even on the Earth however there are life forms that seem wondrously strange to our parochial anthropocentric viewpoint. The recently discovered GFAJ-1 strain of *gammaproteobacteria* for example thrives largely on arsenic, a substance highly toxic to most other terrestrial life forms. This bacterium inhabits Lake Mono in California which naturally has high levels of arsenic and so it has clearly evolved to take advantage of that situation. Hydrothermal vents in the floors of the oceans pour out huge quantities of sulphur compounds and bacteria have evolved to utilize hydrogen sulphide as their energy and food source. In these bacteria chemosynthesis has replaced photosynthesis. The *Clostridium* group of bacteria, which includes those responsible for tetanus and botulism, are strict anaerobes surviving without using oxygen – indeed oxygen is poisonous to them. *Methanogens* are another group of anaerobic bacteria that inhabit the digestive tracts of some animals, including humans and cattle, converting cellulose into methane and are thus partly responsible for the increase in greenhouse gases in the Earth's atmosphere.

When it comes to alien life we should thus only expect beings with six tentacles and three heads if their environment and lifestyle necessitate such extravagances – perhaps they have to eat several different types of food simultaneously whilst also hanging from alien trees and fending off alien predators. Alien plant life, assuming it to be carbon-based, could take many forms but again terrestrial examples suggest that fitness for purpose will tend to produce similar outcomes. The modern non-flowering tree, Gingko biloba, is practically unchanged from when it first evolved 270 million years ago, yet it has a structure and appearance little different from that of (say) birch trees that evolved 200 million year later. In parts of exoplanets deficient in water supplies we might expect plants to evolve that conserve water and so have thick impervious skins and more-or-less spherical shapes like many terrestrial cacti and succulents, whilst in shallow alien seas, plants equipped with gas bladders analogous to some seaweeds could be a possibility. We should not, of course, expect to find any organisms identical with those on the Earth, but if the exoplanet has a similar mass, temperature, composition, etc., to the Earth (i.e. a twin-Earth) then at least some of the life forms upon it should not differ too greatly from those on the Earth.

Exoplanets that differ in some significant way from the Earth – temperature, gravitational field, atmosphere (or lack of it), surface composition, levels of volcanic and tectonic activity, radioactivity levels, magnetic field, presence or absence of liquids, etc. – must be expected to evolve life forms that do differ significantly from the terrestrial experience. However we may still expect environment and lifestyle to influence evolutionary trends. Thus if the surface gravity of an exoplanet is twice that of the Earth it might be impossible for actively flying organisms like birds to evolve, but that ecological niche might be occupied by creatures that have evolved hot gas balloons. On such a planet six legs (or tentacles) might well be an advantage and the analogues of trees would probably be shorter and squatter with stubby branches. Speculation beyond this point though is best left to the reader's imagination – or to the imaginations of his or her favourite SF writers.

Appendix I
Nomenclature – or – What's in a Name?

Planets, Dwarf Planets and Exoplanets

Thanks to recent deliberations by the International Astronomical Union (IAU), Jupiter (mass 318 times that of the Earth and largely gaseous) and Mercury (mass 0.06 that of the Earth and entirely solid) are both PLANETS. Eris and Pluto (masses 0.003 and 0.002 Earth masses and solid) are DWARF PLANETS, Ganymede and Titan though (masses 0.025 and 0.023 Earth masses and solid) are MOONS or SATELLITES, while Pallas (mass 0.00003 that of the Earth and solid) is a MINOR PLANET or ASTEROID.

The IAU definitions of planets etc., as given in 2006 in its 'Resolution B5' are:

A "PLANET" is a celestial body that
(a) Is in orbit around the Sun
(b) Has sufficient mass for its self-gravity to overcome rigid body forces so that it assumes a hydrostatic equilibrium (nearly round) shape, and
(c) Has cleared the neighbourhood around its orbit

A "DWARF PLANET" is a celestial body that
(a) Is in orbit around the Sun
(b) Has sufficient mass for its self-gravity to overcome rigid body forces so that it assumes a hydrostatic equilibrium (nearly round) shape
(c) Has not cleared the neighbourhood around its orbit, and
(d) Is not a satellite

C. Kitchin, *Exoplanets: Finding, Exploring, and Understanding Alien Worlds*, Astronomers' Universe, DOI 10.1007/978-1-4614-0644-0,
© Springer Science+Business Media, LLC 2012

All other objects, except satellites, orbiting the Sun shall be referred to collectively as "SMALL SOLAR SYSTEM BODIES".

Thus Planets are now the eight objects – Mercury, Venus, Earth, Mars, Jupiter, Saturn, Uranus and Neptune, while Dwarf Planets are the five objects – Ceres, Pluto, Haumea, Makemake, and Eris.

The drawbacks of the IAU definitions for the purposes of this book are obvious – only objects within the solar system can be called planets, dwarf planets, or small solar system bodies (the situation for satellites is less clear). So nobody will EVER discover a planet beyond the solar system because by being beyond the solar system it can no longer be called a planet! This is clearly a nonsensical situation and one which will doubtless be 'officially' clarified in time.

Meanwhile most astronomers accept as a working definition of an exoplanet that its mass must be too small for thermonuclear fusion reactions (for example the combination of four hydrogen nuclei to make one helium nucleus) of any sort to be taking place within it now or to have taken place in the past or will take place in the future. The maximum mass for an exoplanet assuming a similar composition for the elements heavier than helium to that of the Sun and based upon the best of the current models for stellar evolution is about 13 times the mass of Jupiter.

The working definition of the minimum mass for an exoplanet (or dwarf exoplanet) is not yet of much significance, since the least massive exoplanet discovered at the time of writing still has twice the mass of the Earth. However the IAU definitions for the Solar System will probably suffice if ever needed.

The minimum mass for a normal star is about 80 times the mass of Jupiter. Between that limit and the top end of the exoplanets (13 Jupiter masses) lie a group of objects known as Brown Dwarfs that are (or have been or will be) obtaining energy from fusion reactions – but only those involving heavy hydrogen (deuterium) and lithium *not* the 'proper' reactions that produce the Sun's and most stars' energies and which are based upon normal (or light) hydrogen. In 1975, when the existence of this class of objects had been predicted theoretically, but long before any of them had been found, Jill Tarter, now Director of the Center for the Search for Extraterrestrial Intelligence (SETI), suggested that they be called 'Brown Dwarfs.' The justification for the name was that, at the time, the colour of brown dwarfs was not known and because 'brown' is not a primary colour it could be used without pre-empting later results. It is, however, a somewhat unfortunate choice, since brown dwarfs are not brown – they range in colour from red, through deep red to infrared and even microwave in their emissions. However the name is now so widely used that it is unlikely ever to be replaced by something more appropriate.

The differences between large exoplanets and small brown dwarfs and between large brown dwarfs and small stars are minimal in the border-line zones and are probably not very important. There are however definite physical criteria for making the distinction when there is need for it. Thus the largest exoplanets have never undergone fusion reactions; the smallest brown dwarfs have done so. The largest brown dwarfs, likewise, have never supported the fusion of normal hydrogen, though all brown dwarfs have fused heavy hydrogen, and those over about 65 Jupiter masses have fused lithium as well.

Exoplanet Names and Labels

When exoplanets first started to be found, they were given names. Methuselah is the name given to the planet orbiting the pulsar-white dwarf binary PSR B1620-26, Bellerophon that given to 51 Peg's exoplanet and Osiris is HD 209458's planet. As large numbers of exoplanets started to be found, the practice was recognised as being too cumbersome and it was replaced by an adaptation of the system used for naming binary stars. With binary and multiple star systems the primary component (the brighter and/or more massive star) is designated by 'A' following the star's main name (details of the usual ways in which stars' names are derived may be found in most general astronomy books), the secondary by 'B', a third component would be 'C' and so on. Thus we have α Cen A and α Cen B, 40 Eri A, 40 Eri B and 40 Eri C and Sirius A and Sirius B. Brown dwarfs are also labelled using this system.

For exoplanets a commonly used convention is to label exoplanets by the lower case letters b, c, d, ... following the name of the star (which is designated by an 'a' or by no extra label) in the order of their discovery or, for two or more exoplanets discovered simultaneously, in the order of their distance from the star. For example, 55 Cnc is a binary star system about 40 light years away from us with a yellow dwarf primary and a red dwarf secondary. Five exoplanets have now been detected orbiting the yellow dwarf. The nomenclature for this system is thus

55 Cnc A (yellow dwarf star)

55 Cnc B (red dwarf star)

55 Cnc Ab, 55 Cnc Ac, 55 Cnc Ad, 55 Cnc Ae and 55 Cnc Af (the five exoplanets – although this nomenclature strictly follows the rules, it is quite common to see the exoplanets labelled as 55 Cnc b, 55 Cnc c, 55 Cnc d, 55 Cnc e and 55 Cnc f).

Another related system uses the same letters but combined with the instrument that made the discovery and a running number giving the order of that discovery. Thus we have Kepler-4 b which is the first exoplanet discovered by the Kepler spacecraft (Kepler-1 b to Kepler-3 b were already known exoplanets) and similarly from the Optical Gravitational Lensing Experiment (OGLE), the Wide Angle Search for Planets (WASP) and the Convection, Rotation and planetary Transit spacecraft (CoRoT), we have OGLE-TR-10 b, WASP-6 b and CoRoT-7 c.

OGLE can also detect exoplanets via gravitational microlensing events and along with the MOA (Microlensing Observations in Astrophysics) search this results in names for planets such as OGLE-2005-071 L b and MOA-2007-BLG-400-L b. The format is: Search name – Year of the microlensing event – sequential number of the event – L for a lensing-based exoplanet discovery (cf. TR for a transit discovery). When included, the term 'BLG' denotes that the star is in the Milky Way's central galactic BuLGe.

No convention for naming exoplanets existed when, in 1992, the first two were found orbiting the pulsar PSR 1257+12. They were therefore labelled as PSR 1257+12 B and PSR 1257+12 C. Two years later, a third exoplanet discovered closer in towards the pulsar was called PSR 1257+12 A and in 2002 a fourth, as yet unconfirmed, member of the system became PSR 1257+12 D. These names are still in use despite the confusion that arises with binary stars. The labels PSR 1257+12 a, PSR 1257+12 b, etc. though are also to be found.

Greek Alphabet

Since the letters of the Greek alphabet are often used as a part of star and exoplanet names, they are listed here for convenient reference.

Letter	Lower case	Upper case
Alpha	α	A
Beta	β	B
Gamma	γ	Γ
Delta	δ	Δ
Epsilon	ε	E
Zeta	ζ	Z
Eta	η	H
Theta	θ	Θ
Iota	ι	I
Kappa	κ	K
Lambda	λ	Λ
Mu	μ	M
Nu	ν	N
Xi	ξ	Ξ
Omicron	o	O
Pi	π	Π
Rho	ρ	P
Sigma	σ	Σ
Tau	τ	T
Upsilon	υ	Y
Phi	φ	Φ
Chi	χ	X
Psi	ψ	Ψ
Omega	ω	Ω

Appendix II
Note on Distances, Sizes and Masses, etc.

SI units (*le Système international d'unités*) are now widely, although not universally, used by scientists and engineers as a logical and coherent set of measures.

For many astronomical purposes, the numbers encountered if SI units are used can be inconveniently large (*'astronomical'* in size!). Thus the distance to the nearest star is 40,000,000,000,000,000 m. Mathematically inclined readers can write this as 4×10^{16} m, but not everyone is happy with this 'index notation' and, in any case, it is still a bit cumbersome (however where very large or very small numbers are written out in full their index equivalents are also given in the text).

Distance

Astronomers have therefore developed a number of non-SI units which are more convenient to use. Thus for distance there are the Astronomical Unit (AU – the average distance between the Earth and the Sun), the Light Year (ly – the DISTANCE travelled by light in a year) and the Parsec (pc). Their reasonably exact values are:

1 astronomical unit = 149,600,000,000 m (= 1.496×10^{11} m)
1 light year = 9,460,000,000,000,000 m (= 9.46×10^{15} m)
1 parsec = 30,860,000,000,000,000 m (= 3.086×10^{16} m)

However the distances to astronomical objects are usually not known very accurately, and so in this book, unless higher accuracy is essential, we shall use the easier approximations:

> 1 astronomical unit ≈ 150,000,000,000 m (= 1.5 × 10^{11} m)
> 1 light year ≈ 10,000,000,000,000,000 m (= 1 × 10^{16} m)
> 1 parsec ≈ 31,000,000,000,000,000 m (= 3.1 × 10^{16} m)

and so

> 1 astronomical unit ≈ 0.000,015 light years ≈ 0.000,005 parsecs
> 1 light year ≈ 0.3 parsecs ≈ 66,000 astronomical units
> 1 parsec ≈ 200,000 astronomical units ≈ 3.3 light years

Giving the distance to the nearest star as 4.3 light years (1.3 parsec, 260,000 astronomical units).

Mass

For the masses of stars and planets, the masses of the Sun, Jupiter and the Earth are convenient units, with:

> Earth mass ≈ 6,000,000,000,000,000,000,000,000 kg (= 6 × 10^{24} kg)
> Jupiter mass ≈ 2,000,000,000,000,000,000,000,000,000 kg (= 2 × 10^{27} kg)
> Solar mass ≈ 2,000,000,000,000,000,000,000,000,000,000 kg (= 2 × 10^{30} kg)

and so

> Earth mass ≈ 0.003 Jupiter masses ≈ 0.000,003 Solar masses
> Jupiter mass ≈ 0.001 Solar masses ≈ 320 Earth masses
> Solar mass ≈ 330,000 Earth masses ≈ 1,000 Jupiter masses

Size

For the sizes of stars and planets, the radii of the Sun, Jupiter and the Earth are similarly convenient measures, with:

> Earth radius ≈ 6,400,000 m
> Jupiter radius ≈ 71,000,000 m
> Solar radius ≈ 700,000,000 m

Giving

> Earth radius ≈ 0.09 Jupiter radii ≈ 0.009 Solar radii
> Jupiter radius ≈ 0.1 Solar radii ≈ 12 Earth radii
> Solar radius ≈ 110 Earth radii ≈ 10 Jupiter radii

Angle

Small angles are measured in minutes of arc (arc-minutes) and seconds of arc (arc-seconds). A minute of arc is a sixtieth of a degree and a second of arc is a 60th of a minute of arc. The symbols 'and "are used for minutes and seconds of arc respectively. Even smaller angles are measured in milli and micro seconds of arc (mas and µas) giving

$1° = 60' = 3,600'' = 3,600,000$ mas $= 3,600,000,000$ µas

At the surface of the Earth (relative to the centre of the Earth), these angles correspond to distances of

$1° = 111.3$ km
$1' = 1.855$ km
$1'' = 30.9$ m
1 mas $= 30.9$ mm
1 µas $= 0.03$ mm

Wavelength and Frequency

Light (and radio waves, microwaves, infrared radiation, ultra-violet radiation, x-ray and gamma rays – which are all different forms of electromagnetic radiation) behaves at times like a wave and at other times as a collection of particles. Readers intrigued as to how this is possible are referred to books and internet sites describing Quantum Mechanics for further information. For the purposes of this book it suffices to know that when light or any of the other forms of electromagnetic radiation is behaving as a wave it is described by its wavelength (the distance between two successive wave peaks or troughs) or by its frequency (the number of wave peaks or troughs that pass a given point in space during one second of time). The relationship between wavelength (symbol, λ, units – metres) and frequency (symbol, ν, units – hertz or cycles per second) is given by the speed of light (symbol, c, and which equals 300,000,000 m/s) as follows:

$$\lambda\nu = c$$

The different types of electromagnetic radiation have roughly the following wavelength and frequency ranges:-

	Wavelength	Frequency
Gamma rays	0–0.001 nm	∞ to 300 EHz $(3 \times 10^{20}$ Hz)
X- rays	0.001–10 nm	300 EHz to 30 PHz $(3 \times 10^{16}$ Hz)
Ultra-violet radiation	10–370 nm	30 PHz to 800 THz $(8 \times 10^{14}$ Hz)
Light (visible radiation)	370–700 nm	800 THz to 400 THz $(4 \times 10^{14}$ Hz)
Infrared radiation	700 nm to 0.1 mm	400 THz to 3 THz $(3 \times 10^{12}$ Hz)
Microwave radiation	0.1–10 mm	3 THz to 30 GHz $(3 \times 10^{10}$ Hz)
Radio waves	10 mm to ∞	30 GHz to 0

The units of Ångstroms (symbol, Å, 1 nm = 10 Å) may also be encountered at times, especially for the wavelengths of visible and infrared radiation.

Appendix III
Further Reading

A small selection of recently available resources for further reading and research.

Web Sites

AAVSO – Transit programme	http://www.aavso.org/observing/programs/ccd/transitsearch.shtml
British interplanetary society	http://www.bis-spaceflight.com/
Centre de Données astronomiques de Strasbourg	http://cdsweb.u-strasbg.fr/
Citizen science project	http://citizensciencealliance.org/projects.html
CoRoT archive	http://idoc-corot.ias.u-psud.fr/
Exoplanet catalog	http://www.planetary.org/exoplanets/
Exoplanet data explorer	http://exoplanets.org/
Exoplanet transit database (ETD)	http://var2.astro.cz/ETD/
EXOTIME	http://www.na.astro.it/~silvotti/exotime/
Extrasolar planets encyclopaedia	http://exoplanet.eu/, http://exoplanet.eu/searches.php (Current exoplanet searches)
Greg laughlin's exoplanet web log	http://oklo.org/
Kepler spacecraft archive data	http://archive.stsci.edu/kepler/data_search/search.php
Known planetary systems	http://www.princeton.edu/~willman/planetary_systems/
MicroFUN-PLANET collaboration	http://planet.iap.fr/
Microlensing follow-up network (MicroFUN)	http://www.astronomy.ohio-state.edu/~microfun/
NStED – NASA/IPAC/NexSci Star and exoplanet database	http://nsted.ipac.caltech.edu/index.html
OGLE archive data – photometry and the star catalogue	http://ogle.astrouw.edu.pl/
Planet hunters	http://kepler.nasa.gov/education/planethunters/

PlanetQuest	http://planetquest.jpl.nasa.gov/index.cfm, http://planetquest.jpl.nasa.gov/links/Search_Map/searches_index.cfm (Current exoplanet searches)
SETI	http://www.seti.org/, http://en.wikipedia.org/wiki/SETI
Spectrashift	http://www.spectrashift.com/index.shtml
SuperWASP public archive	http://www.wasp.le.ac.uk/public/
Tau zero foundation	http://www.tauzero.aero/
Transitsearch	http://www.transitsearch.org/

Magazines and Journals

Astronomical Journal
Astronomy
Astronomy and Astrophysics
Astronomy Now
Astrophysical Journal
Ciel et Espace
Icarus
Monthly Notices of the Royal Astronomical Society
Nature
New Scientist
Publications of the Astronomical Society of the Pacific
Science
Scientific American
Sky and Telescope
Solar Physics

Books

Exoplanets and Alien Life

Aime C., Vakili F. *Direct Imaging of Exoplanets (IAU C200): Science and Techniques (Proceedings of the International Astronomical Union Symposia and Colloquia)* 2006 (Cambridge University press)

Ballesteros F.J. *E.T. Talk: How Will We Communicate with Intelligent Life on Other Worlds?* 2010 (Springer)

Barnes R. *Formation and Evolution of Exoplanets* 2010 (Wiley)

Campanella G. The *search for exomoons and the characterization of exoplanets: Are we alone in the Universe?* 2009 (Lambert Academic Publishing)

Casoli F., Encrenaz T. *The New Worlds · Extrasolar Planets* 2007 (Springer)

Cole G.H.A. *Wandering Stars: About Planets and Exo-Planets, An Introductory Notebook* 2006 (Imperial College press)

Davies P. *The Eerie Silence: Are We Alone in the Universe?* 2010 (Allen Lane)

Dvorak R. *Extrasolar Planets: Formation, Detection and Dynamics* 2007 (Wiley)

Haswell C.A. *Transiting Exoplanets* 2010 (Cambridge University press)

Jones B.W. *Life in the Solar System and Beyond* 2004 (Praxis Publishing)

Jones B.W. *The Search for Life Continued: Planets Around Other Stars* 2008 (Springer Praxis Books)

Kasting J. *How to Find a Habitable Planet* 2009 (Princeton University press)

Mason J.W. *Exoplanets · Detection, Formation, Properties, Habitability* (Springer) 2008

Mayor M. et al. *New Worlds in the Cosmos: The Discovery of Exoplanets* 2003 (Cambridge University press)

Ollivier et al. *Planetary Systems · Detection, Formation and Habitability of Extrasolar Planets* 2009 (Springer)

Seager S. *Exoplanets* 2011 (University of Arizona press)

Shuch H.P. *Searching for Extraterrestrial Intelligence: SETI Past, Present, and Future* 2010 (Springer)

Steves B. et al. *Extra Solar Planets: The Detection, Formation, Evolution and Dynamics of Planetary Systems* 2010 (Taylor and Francis)

Vazquez M. et al. *The Earth as a Distant Planet: A Rosetta Stone for the Search of Earth-Like Worlds* 2010 (Springer)

Introductory Astronomy Books

Freedman R A Kaufmann III W J *Universe* 2001 (WH Freeman)

Inglis M *Astrophysics is Easy!: An Introduction for the Amateur Astronomer* 2005 (Springer)

Montenbruck O, Pfleger T *Astronomy on the Personal Computer* 2000 (Springer)

Moore P, Watson J *Astronomy with a Budget Telescope* 2002 (Springer)

Nicolson I *Dark Side of the Universe* 2007 (Canopus Publishing)

Tonkin S F *AstroFAQs* 2000 (Springer)

Practical Astronomy Books

Dunlop S, Tirion W *Practical Astronomy* 2006 (Philip's)

Gary B. *Exoplanet Observing for Amateurs* 2007 (Reductionist Publications) – The first edition is available to down-load free of charge from http://brucegary.net/book_EOA/x.htm.

Gainer M *Real Astronomy with Small Telescopes: Step-by-Step Activities for Discovery* 2006 (Springer)

Hearnshaw J B *Measurement of Starlight* 1996 (Cambridge University Press)

Howell S B *Handbook of CCD Astronomy* 2000 (Cambridge University Press)

Kitchin C R *Astrophysical Techniques (5th Edition)* 2009 (Taylor and Francis)

Kitchin C R *Optical Astronomical Spectroscopy* 1995 (Institute of Physics Publishing)

Kitchin C R *Telescope and Techniques (2nd Edition)* 2003 (Springer)

Rieke G H *Detection of Light: From the Ultraviolet to the Submillimeter* 2002 (Cambridge University Press)

Robinson K *Spectroscopy, the Key to the Stars: Reading the Lines in Stellar Spectra* 2007 (Springer)

Stuart A *CCD Astrophotography: High Quality Imaging from the Suburbs* 2006 (Springer)

Tonkin S F *Practical Amateur Spectroscopy* 2002 (Springer)

Warner B *A Practical Guide to Lightcurve Photometry and Analysis* 2006 (Springer)

Reference Books

Allen C W *Allen's Astrophysical Quantities* 2001 (Springer)

Kitchin C R *Illustrated Dictionary of Practical Astronomy* 2002 (Springer)

Murdin P (Ed) *Encyclopaedia of Astronomy and Astrophysics* 2001 (Nature and IoP Publishing)

Ridpath I *Oxford Dictionary of Astronomy* 1997 (Oxford University Press)

Shirley J H Fairbridge R W *Encyclopaedia of Planetary Sciences* 2000 (Kluwer Academic Publishers)

Woodruff J (Ed.) *Philip's Astronomy Dictionary* 2005 (Philip's)

Appendix IV
Technical Background – Some of the Mathematics and Physics Involved in the Study of Exoplanets

The Spectral and Luminosity Classification of Stars

The spectral type depends mainly upon the star's surface temperature and for historical reasons is labelled by upper case letters in the somewhat awkward sequence (O being the hottest stars and M the coolest):-

$$O - B - A - F - G - K - M$$

(The mnemonic 'Oh Be A Fine Girl/Guy Kiss Me' may be a help).

Recently the letters L, T and Y have been added onto the end of the sequence to cover very, very cool stars and brown dwarfs, although an unequivocal example of a class Y brown dwarf has yet to be found. Each of the above main spectral classes is sub-divided into ten (with some minor complications) and the sub-division denoted by an Arabic numeral. The temperatures and other data for each spectral class are given below. Other classes which may occasionally be encountered and which denote special cases or superseded groups include W, C, R, N and S.

Stellar surface temperatures, the balance of their radiation emissions and the total luminosities for main sequence stars.

Spectral class	Surface temperature (°C)	Radiation balance (% of total luminosity)			Luminosity relative to the Sun (i.e. – Solar luminosity units – 400,000,000,000,000,000,000,000 W = 4 × 10²⁶ W) for main sequence stars
		UV	Visible	Near IR	
True stars					
O	>50,000–30,000	97	2.5	0.5	>15,000–3,300
B	30.000–9,500	84	13	3	3,300–45
A	9,500–7,000	33	40	27	45–7
F	7,000–5,700	16	42	42	7–1.4
G	5,700–4,900	9	37	54	1.4–0.36
K	4,900–3,600	3	38	69	0.36–0.025
M	3,600–<2,400	0.5	14.5	85	0.025–<0.000,04
True stars and some brown dwarfs					
L	1,700–1,000	0	0.1	99.9	<0.000,01
Brown dwarfs					
T	1,000–400	0	<0.01	99.99	Not applicable
Y	<300	0	0	100	Not applicable

The luminosity class is so called because if two stars have the same surface temperatures but different sizes, then the larger star will be the more luminous – but a better name might be 'star-type class.' The luminosity class is denoted by a Roman numeral from I to VII (again with a few complications not dealt with here) as follows:-

Stellar Luminosity Classes

Luminosity class	Star type	Luminosity relative to the Sun (i.e. – Solar luminosity units – 400,000,000,000,000,000,000,000 W = 4 × 10²⁶ W)
I	Supergiants	50,000–15,000
II	Bright giants	10,000–500
III	Giants	5,000–50
IV	Subgiants	100–5
V	Dwarf or Main sequence	15,000–0.000,001
VI	Sub dwarfs	100–0.001
VII	White dwarfs	0.01–0.000,01

In use, the spectral class is followed by the luminosity class, so that the Sun, for example is G2 V and α Ori (Betelgeuse) is M2 I.

Exoplanet Mass Determination from Radial Velocity Measurements

Symbols and quantities used in this section:-

M_{Planet} is the mass of the exoplanet (kg)
G is the universal gravitational constant (6.67×10^{-11} m³/kg/s²)
P is the orbital period of the star (and exoplanet) (s)
M_{Star} is the mass of the star (kg)
V_{Star} is the maximum or minimum orbital velocity of the star (half the amplitude of the velocity curve) (m/s).

When the radial velocity curve of the star due to its motions induced by an orbiting exoplanet has been found, the mass of the exoplanet is given by:

$$M_{Planet} = \left(\frac{1}{2\pi\,G}\right)^{1/3} P^{1/3} M_{Star}^{2/3} V_{Star}$$

When the orbital plane of the star and exoplanet is inclined to the plane of the sky by an angle, i, the true value of V_{Star}, $V_{Star,True}$, is related to the observed value, $V_{Star,Observed}$, by:

$$V_{Star,Observed} = V_{Star,True}\sin i$$

The masses are therefore also related by a similar equation:

$$M_{Planet,Observed} = M_{Planet,True}\sin i$$

Thus

$$M_{Planet,True} \geq M_{Planet,Observed}$$

For this reason the observed masses of exoplanets are often listed as M_{Planet} sin i (or just as M sin i).

It is often possible to determine the inclination of the host star's rotational axis through spectroscopic measurements of its projected rotational velocity and from knowing its rotational period and radius. If the planetary orbit is assumed to lie in the same plane as the star's equator (like the Sun and the solar system planets), then sin i may be calculated and the actual mass of the exoplanet determined. However a recent study of transiting exoplanets has shown that around a third of such systems have a significant misalignment between the star's spin and the exoplanet's orbit. The misalignments occurred, though, for massive stars (1.2–1.5 solar masses) with large planets. The use of the star's rotation axis inclination to determine that of the exoplanet's orbit is thus probably justified for the lower-mass host stars, but is risky for more massive stars.

Data from Exoplanet Transits

Symbols and quantities used in this section:-

M_{Planet} is the mass of the exoplanet (kg)
M_{Star} is the mass of the star (kg – estimated from its spectral type)
R_{Star} is the star's radius (m – estimated from the star's spectral type)
P is the orbital period of both star and planet (s)
T_A is the amplitude of the transit variation expressed as a fraction of the star's brightness outside of the transit (Fig. IV.1).
T_D is the duration of the transit in seconds (Fig. IV.1).
T_O is the offset of the transit from passing exactly across a diameter of its star as a fraction of the stellar radius (Fig. IV.1)
a is the radius (strictly the semi-major axis) of the exoplanet's orbit in metres
i is the inclination of the orbital plane to the plane of the sky
G is the universal gravitational constant (6.67×10^{-11} m³/kg/s²)

1. Size of the exoplanet's orbit
　　From the timing between successive transits, we may determine the orbital period, P, of the exoplanet. The accurate form of Kepler's third law of planetary motion, derived from Newton's law of gravity, then gives us

$$a = \left(\frac{G(M_{Star} + M_{Planet})}{4\pi^2} \right)^{1/3} P^{2/3}$$

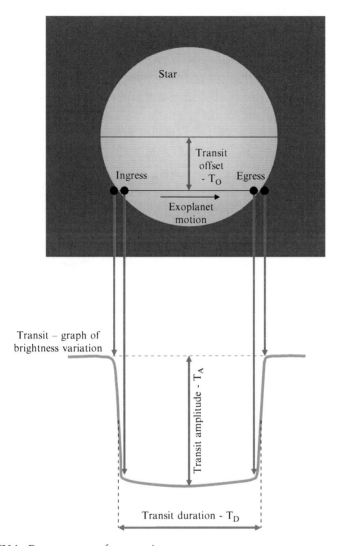

FIGURE IV.1 Parameters of a transit.

However, even for the largest exoplanets, the planet's mass is negligible compared with that of its host star, so the equation becomes

$$a \approx \left(\frac{G M_{Star}}{4\pi^{2}} \right)^{1/3} P^{2/3}$$

thus giving the size of the orbit.

2. Orbital period

When only a single transit has been observed an approximate value for the orbital period may be found by assuming that the exoplanet passes across the centre of the star's disk during the transit, that the orbit is circular and that the star's radius and mass can be estimated from its spectral type. The period is then given by

$$P \approx \frac{\pi \, G M_{Star} \, T_D^3}{4 R_{Star}^3}$$

It should be noted though, that the assumptions behind this method, if unjustified, lead to large inaccuracies in the estimated period

3. Size of the exoplanet

Assuming that the whole of the disk of the exoplanet is silhouetted against the surface of its host star at the centre of the transit and that the star's limb darkening is small, then the planet's radius is given by

$$R_{Planet} = T_A^{1/2} R_{Star}$$

4. Inclination of the orbit

$$Cos \, i = \left(\frac{R_{Star}^2}{a^2} - \frac{\pi^2 T_D^2}{P^2} \right)^{1/2}$$

5. Transit offset

The offset of the transit from passing exactly across a diameter of its star, T_O, is given as a fraction of the stellar radius by

$$T_O = \frac{a \cos i}{R_{Star}}$$

Adaptive Optics

Turbulence in the Earth's atmosphere causes the images observed through it to be distorted – an effect that we see with the naked eye as the twinkling or scintillation of the stars. Images obtained using ground-based telescopes are thus blurred. Typically the blurred image is from 0.5 to 5 (or worse) seconds of arc across. Even small telescopes have much better resolutions than this – an amateur's 0.2-m telescope for example has a theoretical resolution in the visible of about 0.6 arc seconds and so will only perform to approaching its limits from the best observing sites. A 1-m telescope likewise has a theoretical visual resolution of 0.12 s of arc (120 milli-arc seconds) and a 10-m telescope one of 12 milli-arc seconds. In the near infrared the theoretical resolutions are about a factor of four poorer than these values. The blurring effect of the atmosphere is the reason why the comparatively small 2.4-m Hubble space telescope, being above the atmosphere, produces sharper images than much larger ground-based telescopes.

Adaptive optics is an optical system that enables ground-based telescopes to perform at close to their theoretical limits. A small fraction of the light gathered by the telescope is diverted to the adaptive optics system and the distortions produced by the atmosphere are measured about a 1,000 times per second. A small mirror within the light path from the telescope then has its shape altered by a few microns in such a way that it introduces distortions into the light beam that are equal and opposite to those caused by the atmosphere. The final image thus has most of the atmospheric blurring removed.

In order to measure the atmospheric distortions an artificial star is needed (a real one would be even better but is rarely to be found where it is wanted). The artificial star is produced at an altitude of about 90 km within the Earth's atmosphere by shining a powerful laser upwards and it can be positioned close (within a few arc seconds) to the object that is to be observed so that both are in the telescope's field of view.

The use of narrow band filters has recently been pioneered and results in almost completely un-blurred images. However so little light get through such filters that the technique is only suitable for the largest telescopes. So far the only such transiting exoplanets observations have utilised the 10.4- m Gran Telescopio Canarias (GTC).

Coronagraphs

The coronagraph was invented in 1930 by Bernard Lyot to enable observations of the solar corona to be made at times other than during solar eclipses. The solar corona is only a millionth as bright as the solar surface and so is normally hidden in the glare of scattered light from the photosphere. The coronagraph produces an artificial eclipse by obscuring the photosphere with an opaque disk. Though simple in concept, making a workable solar coronagraph requires very careful design.

The stellar coronagraph is similarly designed to enable faint objects to be detected that are hidden in the glare from much brighter ones, such as exoplanets near to stars. In the simple stellar coronagraph, the star's image is obscured by an opaque disk that is just larger than the star's image and which is placed at the telescope's focus. Some light is still diffracted at the edges of the disk and this is eliminated by an aperture (called the Lyot stop) further down the optical path. For observing structures such as proto-planetary disks around stars, the disk may be made opaque only at its centre and to vary smoothly in transparency, becoming clear at its edges.

The stop may be placed outside the telescope (so that it is more like a real eclipse). The stop then often has a complex shape with spikes or leaf-shapes on its edge (like a sunflower) in order to reduce diffraction effects. Several suggestions for spacecraft using such external stops have been proposed.

One recent development is the use of an optical vortex for the stop. This is constructed like a small glass spiral staircase and eliminates the star's light through interference effects. It is likely to allow much smaller telescopes than those used to date to image exoplanets directly. Another recent development for near infrared coronagraphs uses a phase plate. The phase plate is a disk of zinc selenide with annular zones of differing thicknesses. Interference effects arising from the different times that it takes the infrared light to pass through the different thicknesses of the material again lead to the suppression of the star's light. A phase plate commissioned for the VLT's NACO instrument has recently been used to image β Pic b.

The Simultaneous Differential Imager (SDI) is a device that fulfils a similar function to a coronagraph by a different approach. The image is split into four and passed through narrow band filters centred on and just outside the strong methane absorption band in the near infrared. Jovian type exoplanets contain methane and will be fainter when seen through the filters within the absorption region than when compared with the image through filters outside that region. The host star though will have the same brightness in all images. Subtracting one image from another will eliminate the star's image, but leave that of the planet. Exoplanets have yet to be observed using the SDI, but brown dwarfs have been detected.

Gravitational Lensing and Microlensing

Gravitational Lensing

When an asteroid passes close to the Earth its path is bent by the Earth's gravitational field, even though it does not move in a closed orbit around the Earth. In a similar way any two objects flying past each other in space will follow curved trajectories, whether they be dust particles, rocks, comets, moons, planets, stars or galaxies. What is generally less well-known is that the path of a ray of light is also bent as it passes through a gravitational field. The bending of the path of a beam of light is very small compared with that of an asteroid passing the Earth – just 1.75 s of arc for a beam of light (or radio waves, x-rays etc.) that skims the surface of the Sun. The deflection decreases as the light path gets further away from the object – 0.8 s of arc for a beam of light passing one solar radius (700,000 km) away from the surface of the Sun, 0.6 s of arc for one at two solar radii etc. The deflection also decreases as the mass of the object gets smaller. Thus a beam of light just skimming the cloud tops of Jupiter would be deflected by 17 milli arc-seconds and one just skimming the Earth by 0.7 milli arc-seconds.

The angular deflections of a light beam due to gravity are very small, but over astronomical distances their effect can be significant. Thus the 1.75 s of arc deflection of a light beam skimming the Sun would result in that light arriving in Madrid instead of London or Washington instead of Miami. The deflection by a Jupiter-twin planet some 100 light years away from us would be by 100 million kilometres.

A converging lens similarly deflects light and is able to produce an image of the original object because the amount of the deflection *increases* away from the centre of the light beam (Fig. IV.2). The deflection by the gravitational field of a material object though, as we have just seen, *decreases* away from the centre of the light beam. If we were to make a 'lens' that behaved in a similar way to gravitational fields, then it would be shaped like

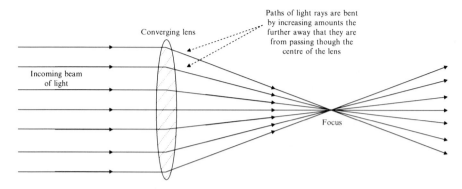

FIGURE IV.2 Optical paths in a conventional converging lens.

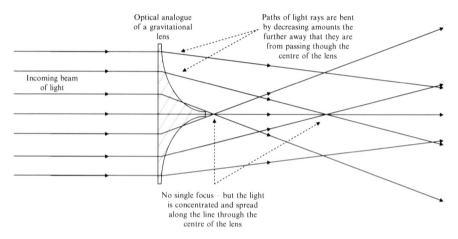

FIGURE IV.3 Optical paths in the optical analogue of a gravitational lens.

the base and stem of a wine glass (Fig. IV.3). The gravitational lens thus does not produce true images of its sources.

Although the gravitational lens does not produce a true image, it does produce optical effects that can be observed. Thus if, say, two stars were to be aligned on almost exactly the same line of sight from the Earth, then the nearer star would be seen to be surrounded by a circle of diffuse light, known as an Einstein ring, that had been 'gravitationally lensed' from the more distant star. The diameter of the Einstein ring is typically around one milli arc-second for the stars whose exoplanets have been observed by

FIGURE IV.4 Caustics produce by reflection from the inner rim of a metal tankard. The shape produced in this common example is called a nephroid cusp (Copyright © C. R. Kitchin 2010).

this method. If the two stars were not quite so exactly aligned as seen from Earth, then the Einstein ring would break up into two spots of light. With more complex structured lensing objects such as galaxies or clusters of galaxies, one, two, three or more spots of light may result from the single more distant source and be separated from each other by tens of seconds of arc. Thus the cluster of galaxies in Leo Minor, known as SDSS J1004+4112, which is 7,000 million light years away from us, appears to have five identical quasars in its midst. In fact the five 'quasars' all originate from an even more distant quasar whose light has been broken up into five spots by the cluster of galaxies itself acting as a gravitational lens.

The optical phenomena displayed by the 'wine glass lens' and related systems can produce concentrations of light where several light paths converge, although these are not true images. The patterns produced in such cases are called optical caustics and often result in cusps such as that shown in Fig. IV.4. Another example likely to be familiar to many people is the pattern of light and shade on the bottom of a swimming pool produced by the waves on the water's surface.

Gravitational Microlensing

As the mass of the lensing objects gets smaller, there comes a point at which the distortion of the source can no longer be perceived, nonetheless the lens and source's combined apparent brightness may still be changed from that in the unlensed state. This effect is termed microlensing. Sometimes the lensing object in these situations is unseen so that only the source is visible. Often also the source and lensing object are small (stars or planets) so that their own motions and/or that of the Earth change the mutual alignment significantly in a short (minutes to years) space of time. The degree of microlensing then alters on the same time scale and the brightness of the source may be seen to be changing. Such a change in the observed brightness is termed a microlensing event. For the simple case of one single star gravitationally lensing another single star, the observed brightness will just peak smoothly (Fig. IV.5a). An event is termed High Magnification when the lens and source are nearly perfectly aligned at their central passage. Brightness increases by up to a factor of a 1,000 are then possible.

Once a third body is involved, such as an exoplanet orbiting the lensing star, there will be caustics produced in the pattern of light received by the observer. The caustics will generally be small compared with the physical sizes of the stars and so the observed variations in brightness will depend greatly upon the exact relative paths of the lens and the source. Generally though, the smooth peak of the pair of single stars will have brightness spikes or other deviations superimposed over a small part of its variation (Fig. IV.5b). Even Earth-sized exoplanets could produce brightness deviations by factors of two or three and Jupiter-sized planets could cause changes by factors of ten or more (cf. changes by 1% or less for exoplanetary transits). Detailed computer modelling of the microlensing event is needed once it is over to give the parameters of the exoplanet such as its mass and orbit.

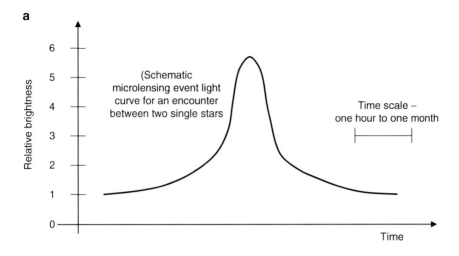

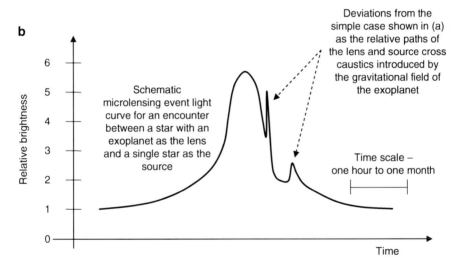

FIGURE IV.5 Schematic light curves for microlensing events. (**a**) An encounter between a single star as lens and a single star as source. (**b**) An encounter between a star with an exoplanet as lens and a single star as source.

Gravitational Microlensing – Sorting out the Data

Targeting regions such as the centre of the galaxy creates major observational problems. Even from good observing sites the stars' images (seeing disks) are up to a second of arc across. In highly crowded fields therefore the images overlap and each pixel in the

CCD may be receiving light from several stars and the total radiation contributing to a single image may originate from millions of individual stars. It is thus impossible to monitor the brightness of each star in the field of view on an individual basis. Instead a 'shotgun' approach to analyzing the data called Difference Image Analysis (DIA) is adopted.

In DIA the highest quality image available of the region under study is first selected to act as the reference image. From this image some 200 stars are chosen that are both bright and not overlapped by other stars. These stars are used to match a newly obtained image of the field to the reference image. When the two images are matched, the reference image is subtracted from the new one. The result, if no stars have changed in brightness and there were to be no noise (which is never the case), should be a uniformly grey 'image.' Where a star has changed in brightness there will be a black or white spot within the grey background (depending upon whether the star has brightened or faded between the two images). In practice, of course, this processing takes place in a computer and so variables will be identified as positive or negative signals within the data set, not as 'black or white spots.'

There will still be numerous stars on the subtracted image because many will be intrinsically variable stars and others will have appeared to have changed because of noise, but there will be far fewer than on the original image. A high proportion of the stars whose variability is due to noise can be eliminated by repeating the process at a different wavelength (say in the blue if the first image was in the red) and then deleting the spurious images which appear at one wavelength but not at the other. Finally the stars whose brightnesses have genuinely altered are sorted and classified to isolate the ones that might be due to microlensing events.

The Jeans' Mass

The Jeans' mass is the minimum mass for an inter-stellar gas cloud to collapse under its own gravitational forces. It is named for Sir James Jeans who was the first to derive the equation, not for the ubiquitous lower body garment. The Jeans' mass should only be taken as an approximate though useful guide to the mass required for the collapse of a gas cloud. The value of the Jeans' mass, in units of the mass of the Sun is given by

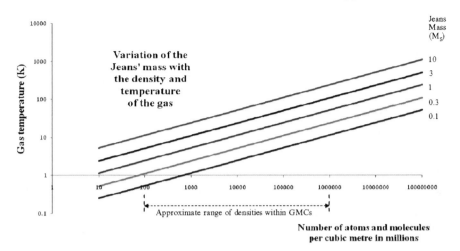

FIGURE IV.6 The variation of Jeans' mass with the temperature (degrees Kelvin) and number density (in millions of particles per cubic metre) of an inter-stellar gas cloud.

$$Jeans'\,mass = 2,600\sqrt{\left(\frac{T^3}{N}\right)}\,Solar\,mass$$

where T is the temperature of the gas in degrees Kelvin (sometimes also called the absolute temperature – add 273.15 to the figure for the temperature in degrees Celsius or Centigrade) and N is the number of gas particles (atoms and molecule) in each cubic metre of the gas.

The variation of the Jeans' mass with the temperature and the number of particles per cubic metre is shown in Fig. IV.6. From there we may see that a mass of gas equal to the mass of the Sun can start to undergo gravitational collapse at a temperature of 3 K (−270°C) at the typical density of a GMC of 100 million particles per cubic metre. However for the dense molecular cores, whose density may reach 1,000,000 million particles per cubic metre, the collapse can begin even when the temperature is as high as ~50°K (around −220°C). Now without very special circumstances (such as inside a low-temperature physics laboratory) temperatures lower than 2.7°K (−270.5°C) cannot be reached because the resid-ual radiation left over from the Big Bang and known as the micro-wave background radiation pervades the whole universe and has a temperature of 2.7°K. Thus it is unlikely that the temperature throughout most of a GMC could fall to low enough values for gravitational collapse to start and in fact their temperatures are usually 10–20°K (–263°C to –253°C).

Within the dense molecular cores however where the density is much higher and which may result from turbulence, random motions or the pressure from nearby stars, collapse will easily be initiated once their mass approaches that of the Sun. Once the collapse starts, both the density and the temperature will rise inside the dense molecular core. If we start with a dense molecular core about 0.1 light years across containing about one solar mass of material, its density will be about 1,000,000 million particles per cubic metre. However by the time it has collapsed by a factor of ten in size (to 0.01 light years), the density will have risen to 1,000,000,000 million particles per cubic metre and the temperature below which the collapse can continue will have risen to over 500°K (>250°C). Another collapse by a factor of ten in size (to 0.001 light years = 70 astronomical units – comparable with the present size of the solar system) and the temperature would have to exceed 5,300°K (5,000°C) for the rising gas pressure to be able to halt the collapse. Thus the collapse will continue, at least until nuclear fusion reactions start and raise the central temperature to millions of degrees.

More important from the point of view of this book is that the temperature rise will be greatest at the centre of what we may now call the proto-star. Away from the centre the density will have risen but the temperature will be much reduced, perhaps remaining little more than the original few tens of degrees in the outer reaches of the proto-star. The Jeans' mass in those regions will then fall to as little as the mass of Jupiter and within the overall collapse of the proto-star smaller concentrations can start to collapse to form planetary-sized masses.

Population Growth – or Malthus *Will* Be Right (Eventually)

In 1798 the Reverend Thomas Malthus wrote in his *An Essay on the Principle of Population*

> ... the power of population is indefinitely greater than the power in the Earth to provide subsistence for man.

In 1798 the total human population of the world was probably a little under 1,000 million. Malthus would undoubtedly have expected the present world population, well in excess of 6,000 million, to have gone beyond "the power in the Earth to provide subsistence for man." The opening-up of new land for the growth of crops and improvements in agriculture have so far postponed Malthus' vindication to the point that his ideas are now often regarded with derision. However in mathematical terms, the growth of population follows a geometric sequence, the increase in resources follows an arithmetical (or slightly better) sequence. Given time, the value generated by a geometrical sequence (which is represented by an exponential curve) will *always* exceed the value generated by a lower order sequence (represented in this case by linear, quadratic or cubic curves).

The idea that exoplanets, if we could reach them, could provide the space needed for the additional numbers of people once the Earth is 'full' seems to provide a permanent solution to the population growth problem. Unfortunately the power of population growth is merely delayed by such an increase in the resources available. Eventually, to add to Malthus' statement

> ... the power of population is indefinitely greater than the power in the Earth *and any additional planets that may become available* to provide subsistence for man.

and we may illustrate this by some calculations based upon extremely optimistic assumptions.

The level of population supportable by the Earth and any equivalent planet depends upon the expectations of the individuals within that population. Clearly a much larger subsistence-level population would be possible than one in which the expectations are those of the inhabitants of currently developed nations on the Earth. The present world population of over 6,000 million may well already exceed the latter criterion. A subsistence-level maximum population might reach 20,000 million. However a subsistence-level standard of living seems a poor ambition to have for our descendents, so let us take 6,000 million as the maximum population sustainable on an Earth-like planet – it will make little difference to the argument – even taking the maximum as 6,000,000 million would postpone the timings calculated below by less than five centuries.

In order to even start the calculation we need a bit of help.

Suppose firstly that we could instantly convert reasonably Earth-like planets to genuine twin-Earths with the wave of a magician's wand. Suppose also that we have another wizard who provides us with transport at almost the speed of light. Finally suppose that a third wizard ensures that every star possesses an exoplanet that is convertible to a twin-Earth.

Our obvious first step, given those three miracles, would be to convert Venus to a twin-Earth. However in the last few decades our population growth has averaged 1.5% per year. Thus to keep the Earth's human population stable at around 6,000 million people, some 90 million people and all their requirements would have to emigrate to Venus every year. In just 50 years, Venus would also have a population of 6,000 million (taking account of the net population increase amongst the new inhabitants of Venus). To keep the populations of both Venus and the Earth stable, emigrants would then have to go to another planet (say an hypothetical twin-Earth belonging to Proxima Centauri) at a rate of 180 million people per year. Within 25 years that planet would also be filled with 6,000 million people. The next suitable exoplanet would take just 25 years to fill, the one after that 15 years and the one after that 12 years and so on.

By the year 2295 AD (assuming the process started in 2010) a new exoplanet would be required every year and would be filled with people within that year. A 1,000 years from now, with our wizard-gifted ability to travel at nearly the speed of light, we could be opening up some 36,000 twin-Earths per year (four an hour!) – BUT – the increase in the human population by then would be requiring 43,000 new twin-Earths per year.

The passage of another century or so to the year 3127 (1,117 years from now) and all 17 million twin-Earths within a radius of 1,117 light years of the Earth would each be populated by 6,000 million people. The population growth would be requiring the opening up of a quarter of a million new exoplanets per year but less than 50,000 would be becoming available in that same time. The population on all planets from then on must start its inevitable increase to unsustainable levels.

The rather fanciful scenario just outlined (there are currently no signs of the three required wizards making their essential appearances) is extremely optimistic. Making it still more optimistic by assuming that there are ten (say) twin-Earths per star would only put off the crunch point by less than two centuries. More effective would be reducing the rate of population increase – 0.5% per year would delay the crunch until 4,000 years time, while a 0.1% growth rate would postpone the evil day until 28100 AD – which would at least see humankind's expansion reaching the centre of the Milky Way galaxy.

"A-ha" I can hear the sceptics saying "we have proved Malthus wrong so far and we will continue to do so. In a 1,000 years from now we will be able to chop up Jupiters and super-Jupiters and make thousands more twin-Earths than the number that you have assumed".

OK – let us grant the sceptics their point, and indeed, go much further. Let us assume that somehow human beings could learn to exist in isolation – i.e. just as individuals hanging around in a vacuum, not breathing, drinking, eating or using any resources in any fashion. That way we would no longer need the Earth, or twin-Earths or Jupiters or super-Jupiters or even stars, nebulae and galaxies. If we then have a fourth wizard to grant us the ability to travel at any speed we like, then all that material would be available to become human beings so creating the first Humiverse.

The total amount of material in the visible universe (I think we may justifiably end this calculation at the edge of the visible universe and also ignore the hypothetical dark matter and dark energy) is around 10^{52} kg – or 2×10^{50} human beings at 50 kg per person. Starting at 6,000 million people in the year 2010, a growth rate of 1.5% would produce the

20,000,000,000,000,000,000,000,000,000,000,000,000,000,000,000,000

people required for the Humiverse by the year 8277 AD. Even a growth rate of 0.1% would do the job by the year 95364 AD. So in less than 100,000 years from now, perhaps less than 10,000 years, the score line will read Malthus 1 Sceptics 0 and Malthus' quotation may be further modified to what must surely be its final form

... the power of population is indefinitely greater than the power in the *Universe* to provide subsistence for man.

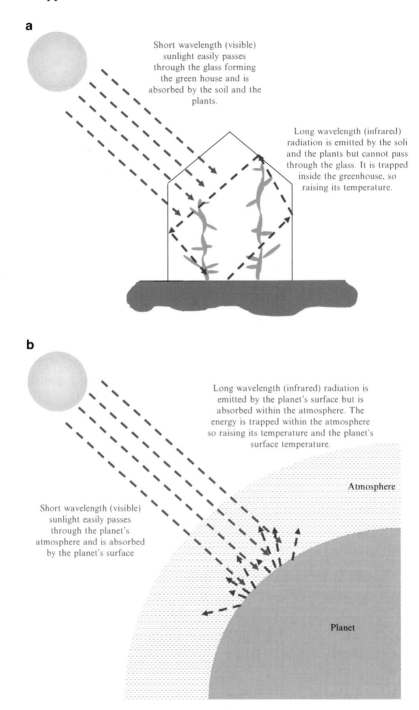

FIGURE IV.7 (*Top*) – The greenhouse effect in a greenhouse. (*Bottom*) – The greenhouse effect on a planet.

Clearly, since we cannot in the foreseeable future even take that first step (colonizing a terra-formed Venus) towards the above schemes of galactic conquest or Humiverse construction, exoplanets do not provide any outlet for population growth except in the very short term.

The Greenhouse Effect

An unheated green house even on a moderately sunny day has a temperature inside that is higher than the temperature of its surroundings. The reason for this is that the glass forming the walls and roof of the greenhouse is transparent to the light coming from the Sun but is opaque to the long-wave radiation emitted by the ground and plants inside the greenhouse. The sunlight thus gets into the greenhouse, is absorbed by the ground and the plants but when that energy is re-emitted it is trapped inside. Not until the temperature has risen quite a bit does a balance between the incoming and out-going energies re-establish itself (Fig. IV.7a).

Several gases such as methane and carbon dioxide have a similar property to that of glass of being transparent to visible radiation but opaque to the infrared. A planet with an atmosphere containing those gases thus has a raised surface temperature compared with a planet without an atmosphere (Fig. IV.7b). The visible light from its star passes through the atmosphere and is absorbed by the planet's surface. The surface then re-radiates this energy but in the infrared. The infrared radiation is absorbed within the atmosphere thus raising both the atmosphere's temperature and the surface temperature of the planet.

The greenhouse effect is not all bad though. Without the warm blanket of its atmosphere the Earth's average temperature would be a very chilly $-18°C$ – instead it is $14°C$. It is thus the *increase* in the greenhouse effect arising from higher levels of carbon dioxide and methane in the Earth's atmosphere that is the worrying problem at the moment, not the greenhouse effect itself.

Appendix V
Names, Acronyms
and Abbreviations

AAPS	Anglo-Australian Planet Search program
AAVSO	American Association of Variable Star Observers
ACCESS	Actively corrected Coronagraph Concepts for Exoplanetary System Studies
AFOE	Advanced Fiber-Optic Echelle
ALMA	Atacama Large Millimeter Array
ANI	Astronomical Nulling Interferometer
APF	Automated Planet Finder
API	Antarctic Plateau Interferometer
APT	Tennessee Automatic Photoelectric Telescope APT
ASP	Arizona Search for Planets
ASPENS	Astrometric Search for Planets Encircling Nearby Stars
ATLAST	Advanced Technology Large-Aperture Space Telescope
AXA	Amateur Exoplanet Archive
BEST II	Berlin Exoplanet Search Telescope II
BOSS	Big Occulting Steerable Satellite
CAPS	Carnegie Astrometric Planet Search
CES	Coude Echelle Spectrometer
CFC	Chlorofluorocarbon compound
CFHT	Canada-France-Hawaii telescope
CHEAP	CHickenfeed Expedition to Another Planet
CIA	Coronographe Interferentiel Achromatique
CNES	Centre National d'Etudes Spatiales
CONICA	Coudé Near Infrared Camera
CoRoT	COnvection, ROtation and planetary Transit
CRIRES	CRyogenic high-resolution InfraRed Echelle Spectrograph (VLT)
DIA	Difference Image Analysis
EDI	Externally Dispersed Interferometry
E-ELT	European Extremely Large Telescope

EPOCh	Extrasolar Planet Observations and Characterization
EPRG	Extrasolar Planets Research Group
ESO	European Southern Observatory
ESPRESSO	Echelle Spectrograph for Rock Exoplanet and Stable Spectro-scopic Observation
ET or ETs	Extra-Terrestrial (intelligences – i.e. alien beings)
ETD	Exoplanet Transit Database
EXPORT	EXo Planetary Observational Research Team
FKSI	Fourier-Kelvin Stellar Interferometer
FLAMES	Fibre Large Area Multi-Element Spectrograph
GEST	Galactic Exoplanet Survey Telescope
GITPO	GIant Transiting Planets Observations
GMAN	Global Microlensing Alert Network
GMC	Giant Molecular Clouds
GMT	Giant Magellan telescope
GPI	Gemini Planet Imager
GTC	Gran Telescopio Canarias
HARPS	High Accuracy Radial velocity Planetary Search
HARPS-NEF	High Accuracy Radial velocity Planetary Search – New Earth Facility
HATNet	Hungarian Automated Telescope Network
HF	Hydrogen Fluoride
HiCIAO	High Contrast Instrument for the Subaru next generation Adaptive Optics
HJD	Heliocentric Julian date
HJD'	Heliocentric Julian date minus 2450000
HSDT	Hubble Space Telescope
HZPF	Habitable Zone Planet Finder
IAC	Instituto de Astrofisica de Canarias
IAU	International Astronomical Union
IKAROS	Interplanetary Kite-craft Accelerated by Radiation Of the Sun
IPAC	Infrared Processing and Analysis Center
IPMO	Isolated Planetary Mass Object
ITASEL	ITAlian Search for Extraterrestrial Life
JWST	James Webb Space Telescope
KBO	Kuiper Belt Object
KVA	Kungliga Vetenskapsakademien
KOI	Kepler Object of Interest
LBT	Large Binocular Telescope
LCOGT	Las Cumbres Observatory Global Telescope
LHC	Large Hadron Collider

LMC	Large Magellanic Cloud
LOFAR	LOw Frequency ARray
MACHO	MAssive Compact Halo Object
MAST	Multi-Mission archive at STScI
METIS	Mid-infrared E-ELT Imager and Spectrograph
MicroFUN	Microlensing Follow-Up Network
MOA	Microlensing Observations in Astrophysics
MONET	MOnitoring NEtwork of Telescopes
MOST	Microvariability and Oscillations os STars
MPS	Microlensing Planet Search project
NACO	NAOS-CONICA
NAOS	Nasmyth Adaptive Optics System (VLT)
NExScI	NASA Exoplanet Science Institute
NICI	Near Infrared Coronagraphic Imager
NICMOS	Near Infrared Camera and Multi-Object Spectrometer
NRAO	National Radio Astronomy Observatory
NStED	NASA/IPAC/NexScI Star and Exoplanet database
OBSS	Origins Billion Star Survey
OGLE	Optical Gravitational Lensing Experiment
OWL	OverWhelmingly Large telescope
PASS	Permanent All Sky Survey
PECO	Pupil mapping Exoplanet Coronagraphic Observer
PHASES	Palomar High precision Astrometric Search for Exoplanet Systems
PICTURE	Planet Imaging Concept Testbed Using a Rocket Experiment
PISCES	Planets In Stellar Clusters Extensive Search
PLANET	Probing Lensing Anomalies NETwork
PLATO	Planetary Transits and Oscillations of stars
PRVS	Precision Radial Velocity Spectrometer
RIPL	Radio Interferometric Planet Search
SDI	Simultaneous differential Imager
SEE	Super Earth Explorer
SETI	Search for ExtraTerrestrial Intelligence
SF	Science Fiction
SIM	Space Interferometry Mission
SKA	Square Kilometer Array
SNDM	Solar Nebular Disk Model
SOFIA	Stratospheric Observatory For Infrared Astronomy
SOPHIE	Spectrographe pour l'Observation des Phénomenes sismologique et Exoplanétaires
SPHERE	Spectro-Polarimetric High-contrast Exoplanet REsearch

SPICA	Space Infrared Telescope for Cosmology and Astrophysics
SPIRIT	SPace InfraRed Interferometric Telescope
STARE	STellar Astrophysics & Research on Exoplanets
STELLA	STELLar Activity
STEPS	STEllar Planet Survey
STEPSS	Survey for Transiting Extrasolar Planets in Stellar Systems
STScI	Space Telescope Science Institute
SuperWASP	Super Wide Angle Search for Planets
TEP	Transits of Extrasolar Planets
THESIS	Terrestrial and Habitable zone Exoplanet Spectroscopy Infrared Spacecraft
TMT	Thirty Meter Telescope
TNO	Trans-Neptunian Object
TPF	Terrestrial Planet Finder
TRAPPIST	Transiting Planets and Planetesimals small telescope
TrES	Transatlantic Exoplanet Survey
TRESCA	TRansiting ExoplanetS and CAndidates
UKIRT	United Kingdom InfraRed Telescope
UMBRAS	Umbral Missions Blocking Radiating Astronomical Sources
UNSWEPS	University of New South Wales ExoPlanet Search
UStAPS	University of St. Andrews Planet Search
UVES	Ultra-Violet Echelle Spectroscope
VIDA	VLTI Imaging with a Densified Array
VLA	Very Large Array
VLT	Very Large Telescope
VLTI	Very Large Telescope Interferometer
WASP	Wide Angle Search for Planets
WHAT	Wise observatory Hungarian made Automated Telescope
ZIMPOL	Zurich Imaging Polarimeter

Index